Global Percussion Innovations

First emerging in North America and Europe in the late 1920s, contemporary percussion practices have transitioned from the fringes of contemporary music to the forefront over the past 90 years. In the 1960s contemporary percussion practices reached Australian shores and a new generation of artists added their voices to this narrative. The role of Australian activity is not yet embedded in the wider narrative of international contemporary percussion, nor is the significance of developments in contemporary percussion practices fully realised in the context of Australian music history. In this monograph, political, social and cultural influences on this art form will be examined for the first time in a historical survey of contemporary percussion music in Australia over a 50-year period, from 1960 to 2010. The rise of the percussion ensemble in the twentieth century to a standard chamber music ensemble is now recognised as one of the major advances in western art music practice internationally. A focus will be placed on ensemble activity via definitive documentation and analysis of ensembles that are amongst the most pioneering and longest established of Australian contemporary music organisations, including the Australian Percussion Ensemble, Synergy Percussion, Adelaide Percussions, Nova Ensemble, Tetrafide Percussion, Taikoz, Clocked Out and Speak Percussion amongst others. Closing with a discussion of influences and identity, this historical narrative will expand our understanding of the impact of Australian contributions to the international contemporary music scene while simultaneously examining how developments in contemporary percussion have contributed to Australia's cultural identity.

Louise Devenish's artistic practice lies at the nexus of performance, research and education in percussion. Key projects include APRA AMC award-winning solo series *Music for One Percussionist*, *The Sound Collectors* duo, Australia's leading percussion group *Speak Percussion* (Vic) and acclaimed electroacoustic sextet *Decibel* (WA), with whom she regularly performs at major festivals throughout Australasia, Europe, North America and the UK. A passionate advocate of new and Australian music, Louise has commissioned over 50 works for percussion, released solo album *music for percussion and electronics* (Tall Poppies) and published in *Contemporary Music Review*, *Percussive Notes* and *Musicology Australia*. Louise is Head of Percussion at the University of Western Australia Conservatorium of Music and is a Churchill Fellow.

Routledge Research in Music Series

Current Directions in Ecomusicology
Music, Culture, Nature
Edited by Aaron S. Allen, Kevin Dawe

Liveness in Modern Music
Musicians, Technology, and the Perception of Performance
Paul Sanden

Vocal Music and Contemporary Identities
Unlimited Voices in East Asia and the West
Edited by Christian Utz, Frederick Lau

Music Video After MTV
Audiovisual Studies, New Media, and Popular Music
Mathias Bonde Korsgaard

Masculinity in Opera
Edited by Philip Purvis

Music, Performance, and the Realities of Film
Shared Concert Experiences in Screen Fiction
Ben Winters

Burma, Kipling and Western Music
The Riff from Mandalay
Andrew Selth

Global Percussion Innovations
The Australian Perspective
Louise Devenish

For more information about this series, please visit: www.routledge.com/music/series/RRM

Global Percussion Innovations
The Australian Perspective

Louise Devenish

Routledge
Taylor & Francis Group
LONDON AND NEW YORK

First published 2019
by Routledge
2 Park Square, Milton Park, Abingdon, Oxon OX14 4RN

and by Routledge
711 Third Avenue, New York, NY 10017

Routledge is an imprint of the Taylor & Francis Group, an informa business

British Library Cataloguing-in-Publication Data
A catalogue record for this book is available from the British Library

Library of Congress Cataloging-in-Publication Data
Names: Devenish, Louise, author
Title: Global percussion innovations : the Australian perspective / Louise Devenish.
Description: Abingdon, Oxon ; New York, NY : Routledge, 2019. | Series: Routledge research in music | Includes bibliographical references and index.
Identifiers: LCCN 2018020021 | ISBN 9781138745612 (hardback) | ISBN 9781315180595 (ebook)
Subjects: LCSH: Percussion ensembles—Australia—20th century—History and criticism. | Percussion ensembles (Musical groups)—Australia.
Classification: LCC ML1151.A87 D48 2019 | DDC 786.80994—dc23
LC record available at https://lccn.loc.gov/2018020021

ISBN: 978-1-138-74561-2 (hbk)
ISBN: 978-1-315-18059-5 (ebk)

Typeset in Times New Roman
by Apex CoVantage, LLC

To Alison and Allan Conroy

Contents

Figures

Tables

Musical examples

Abbreviations

ACME	Australian Contemporary Music Ensemble
AETT	Australian Elizabethan Theatre Trust
APE	Australian Percussion Ensemble
ASO	Adelaide Symphony Orchestra
AYO	Australian Youth Orchestra
MSO	Melbourne Symphony Orchestra
NSW	New South Wales
PPO	Percussion Pops Orchestra
SSO	Sydney Symphony Orchestra
TSO	Tasmanian Symphony Orchestra
UWA	University of Western Australia
VCA	Victorian College of the Arts
WA	Western Australia
WASO	West Australian Symphony Orchestra

Acknowledgements

Thank you to the wonderful and generous community of Australian percussionists and their collaborators who shared their work with me over the course of this research: Michael Askill, James Bailey, Dalmazio Babaré, Cherie Cadd, Mark Cain, Ian Cleworth, Timothy Constable, Rob Cossom, Wendy Couch, Amanda Dean, Ross Dovey, Claire Edwardes, Gary France, Jean-Charles François, Fleur Green, Amanda Grigg, Moya Henderson, Gordon Hughes, Dennis Johnson, Rebecca Lagos, Graeme Leak, Robert Lloyd, Kevin Man, Anthony Maydwell, Richard Miller, David Montgomery, Peter Neville, Catherine Oates, Tom O'Kelly, Colin Piper, Daryl Pratt, Ryszard Pusz, David Pye, Barry Quinn, Ron Reeves, Steven Richter, Iain Robbie, Paul Sarcich, Leah Scholes, John Seal, Neville Talbot, Paul Tanner, Alex Timcke, Vanessa Tomlinson, Eugene Ughetti, Gary Wain, Nigel Westlake and Tim White.

Thank you to my friends and colleagues who read drafts, assisted with formatting or offered advice in the preparation of this book: Susan Broomhall, Jonathan McIntosh, Cecilia Sun, David Symons and Christopher Tonkin at the University of Western Australia; Vanessa Tomlinson at Queensland Conservatorium Griffith University and Pedro Alvarez.

Special thanks to my family, who always support my journey in percussion music in so many different ways: Luke Devenish, Alison Conroy and Julia Trezise-Conroy.

Introduction

First emerging in North America and Europe in the 1920s, contemporary percussion practices have transitioned from the fringes of contemporary music to the forefront over the past 90 years. In the 1960s, contemporary percussion practices reached Australian shores and in the years that followed a new generation of artists added their voices to this narrative. A relatively rapid period of development has seen percussion establish itself in Australia as a stylistically diverse and continually evolving discipline. However, the role of Australian activity is not yet embedded in the wider narrative of international contemporary percussion, nor is the significance of developments in contemporary percussion practices fully realised in the context of Australian music history. In this book, political, social and cultural influences on this art form are examined for the first time in a historical survey of contemporary percussion music in Australia over a 50-year period, from 1960 to 2010.

Investigation from a broad perspective allows parallels to be drawn between developments in contemporary percussion practices and changes in the Australian cultural landscape during this 50-year period. This survey expands our understanding of the impact of Australian cultural contributions to the international contemporary music scene while simultaneously examining how developments in contemporary percussion have contributed to the shaping of Australia's cultural identity. Examination of key historical events that effected change are explored in a broader international context, supported with discussion of significant repertoire. Together, these components initiate a discussion surrounding identity and schools of thought in the percussive arts.

The 1960s have been selected as the point of departure for the book, representing the commencement of considerable growth and change in Australian contemporary music. In the 1960s and 1970s, composers and performers were increasing in number, government and independent organisations supporting contemporary music were established and new ensembles began exploring new forms of contemporary music making in chamber music, electronic music and orchestral music. It was within this climate of musical change that contemporary percussion first appeared in Australia. Less than five decades after it first emerged in the 1970s, percussion activity that once existed at the fringes of Australian contemporary music is now amongst the most prominent Australian contemporary music activity. In 1970, there were only a handful of trained percussionists in Australia and no original works for percussion ensemble by Australian composers in existence. Today, contemporary percussion is practised by dozens of professionals in each major city and the repertoire numbers in the thousands. New works for solo percussion and ensembles are commissioned at a growing rate. Virtuosic percussion ensembles presenting increasingly ambitious performance projects are the cornerstone of contemporary percussion activity in most major cities. These factors have contributed to an increased presence of Australian contemporary percussion practices in both the Australian performing arts climate and the international music scene.

Internationally, contemporary percussion is an ever-growing practice characterised at a fundamental level by exploration. Some twentieth-century definitions of percussion reference lists of instruments and objects so long that it would almost be easier to compile a list of the instruments and objects that aren't considered percussive. This has led others to describe percussion instruments both as 'any physical object that can be sounded' (Ughetti, 2017) and a 'non-instrument' (Schick, 2017). These broad definitions suggest that instrumentation is not always the primary feature of percussion music. When a deeper understanding of percussion is required, sounding techniques are often referenced. Initially the primary sounding techniques of striking, shaking or stroking were referenced, however, as the definition of a percussion instrument has grown, so has the range of sounding techniques to include acoustic and electronic means, both amplified and unamplified. Looking beyond instrumentation and technique, percussion is characterised by experimental music processes that draw on influences from world music, improvisation and the visual and theatre arts from a western art music perspective, processes that are increasingly occupying the space between contemporary music and conceptual performance art. Recently, this has shifted some of the focus to personal approach and the role of the performer. Percussion has been described as more of a concept than an instrument (Esler, 2014: 113), and Tomlinson states: 'Far clearer for me, is to think of percussion as an attitude, an investigation of sound . . . I think more of the person *being* a percussionist – a sound maker and explorer' (Tomlinson, 2017).

All of this speaks to the time and place contemporary percussion developed. First emerging in the twentieth century, the discipline of percussion is free of the traditions maintained by other western musical instruments, such as violin and piano. Consequently, many current practitioners refer to contemporary percussion and contemporary music as the music of our time. The rise of the percussion ensemble in the twentieth century to a standard chamber music ensemble is now recognised as one of the major advances in western art music practice internationally, and suggestions that the twenty-first century percussion quartet is as influential as the eighteenth-century string quartet are not uncommon.

Although the rate of development of contemporary percussion performance and education practices internationally has been rapid, scholarly research surrounding contemporary percussion has not developed at the same rate in countries where the art form flourishes. Burgeoning scholarly interest in contemporary percussion in the twentieth century and beyond has been primarily focussed on activity in North America and parts of Europe. Since 2000, only three significant scholarly books in contemporary percussion have been published by practitioner-researchers: a book by Steven Schick (University of Rochester Press, 2006), and edited collections compiled by Kevin Lewis and Gustavo Aguilar (Routledge, 2014) and by Russell Hartenberger (Cambridge University Press, 2016). All three examine the topic primarily from a North American perspective, thereby excluding information on Australian practices or repertoire. Recent book publications examining Australian contemporary music during the same historical period have focussed primarily on more conservative musical genres such as orchestral symphonies, traditional chamber music ensemble repertoire and analysis, experimental and electroacoustic music, or gender-based music studies (for examples, see Hooper, 2012; Rogers, 2009; Kerry, 2009; Symons, 2016; McNeill, 2014; Priest, 2009; Appleby, 2012). None of these publications permitted examination of contemporary percussion. One of the reasons for this exclusion is the genre's diversity and evolving form. Although occupying a space at the forefront of contemporary music activity that draws upon influences from the aforementioned areas and others, contemporary percussion practices occupy a unique 'other' place in music.

Amongst practitioners and musicologists, the relationship between Australian contemporary percussion practices and international contemporary music is understood but undocumented and therefore undeveloped, and this book offers an opportunity to include significant developments in Australian music in the context of international scholarly research in the field at this formative time. That Australian contemporary percussion history has not yet been documented is not particularly surprising, for two reasons, the first of which is the changing nature of percussion, as discussed earlier. Second, Australian contemporary percussion practices began as an extension of practices established in North America and Europe in the early and mid-twentieth century. The relative youth of the discipline on an international level applies not just to creative output but to scholarly output as well.

As global contemporary percussion practices continue to advance, knowledge of the roots of these practices becomes increasingly important. Access to history – and therefore an understanding of the context of new creative work – is an important part of any artistic discipline; this book offers the first systematic documentation of the emergence and development of contemporary percussion practices in Australia. Due to the lack of existing literature, access to Australian percussion history has been limited for most percussionists, usually occurring in the form of informal conversations with older colleagues. A stock of stories exists in each city, as teachers delight in telling students about rehearsing in lifts and corridors at universities before percussion studios existed; building versions of instruments they could not purchase in backyards; and colleagues reminiscing about creating new works with composers that have since become staples of the Australian percussion repertoire. Intermingled with these are many more tales: indeed, telling stories while setting up and packing down immense percussion set-ups is something of a global percussion tradition. Although this has been sufficient as a way of sharing history in Australia for many years, it is accessible only to a small group of people and relies largely on the intangibility of memory. Furthermore, this history can only extend back as far as the memories of living practitioners. In recognition of the importance of first-hand accounts, the methodology for this research involved a combination of interviews, literature surveys, archival research, repertoire analysis and reconstruction.

Emblematic to the development of contemporary percussion practices in Australia, examination of percussion ensemble activity throughout the 1970s, 1980s and 1990s is a focus of the book's narrative. Chapter One includes an exploration of the Australian cultural landscape in the 1960s and 1970s. During these decades, significant cultural change took place across all art forms as Australian artistic culture began to move away from previous influences, primarily from England, towards new approaches. These changes in the cultural climate made room for innovative contemporary percussion practices to emerge. Simultaneously, as artists began to explore their uniquely Australian voices, social changes took place that supported their creative ventures. In addition to a gradual increase in acknowledgement of Australian creative output by audiences and critics, new professional and educational opportunities, the formation of substantial federal funding bodies and establishment of independent organisations inspired a new generation of Australian composers and performers. Dozens of performing arts groups were established, including a number of percussion ensembles. Further, artists of all disciplines began taking new creative risks, resulting in greater diversity of output. During this period, the establishment of links between international and Australian contemporary music commenced with the return of expatriates, touring and immigration of international musicians. The influence of the world's first percussion ensemble – Les Percussions de Strasbourg – on nascent contemporary percussion activity in Australia is featured in this chapter.

Chapters Two through Six document and analyse Australian contemporary percussion activity for the first time in the context of major international developments in the genre and a changing local cultural climate, providing a new lens through which Australian contemporary music history can be viewed. These chapters are organised geographically, as the large distances between resulted in the development of unique approaches to percussive music making in different Australian cities. Chapter Two examines early contemporary percussion activity in Australia through the pioneering activities of the Australian Percussion Ensemble in Melbourne, and discusses the change in percussion activity effected by the establishment of the Victorian College of the Arts in 1974. The chapter concludes with a brief discussion of experimental approaches post-2000 that continued the trajectory of growth commenced in the 1970s. Changes in performance practices in the twenty-first century demonstrate an increased influence of performance art and conceptual art on contemporary percussion projects, particularly in the work of Speak Percussion. Chapter Three continues this narrative in comprehensively documenting the activities of Synergy Percussion in Sydney, Australia's longest-running contemporary music ensemble. Analysis of the impact of this group first on national, then on international contemporaries extends beyond ensembles to solo contributions and interdisciplinary collaboration. This is followed by an examination of contemporary Japanese-Australian drumming ensemble Taikoz and discussion of independent activities by individual artists, including case studies of significant projects by Graeme Leak. Chapters Four and Five look to percussion activity in South Australia and Western Australia respectively, examining the impact of the relative geographical isolation of these practitioners from their contemporaries in Europe and North America. Unique outputs blending contemporary western and traditional musics from neighbouring countries feature here. Chapter Six concludes the narrative focussed on percussion ensemble activity with examinations of nascent practices and individual activities, providing a broader cross section of percussion activity. Here, work in Brisbane, Canberra and Hobart is explored.

The final chapter is a key component of the monograph, reflecting on questions of identity and influence that arise in previous chapters. Are there any defining characteristics to be found in the body of Australian percussion music? If there are, are these characteristics unique to Australian percussion? How have Australian practices aligned with international practices? How have advances in technology affected the development of global schools of thought? The nature of percussion itself renders succinct answers to these questions difficult; on a global scale, we can see that even the definition of a western percussion instrument is constantly changing. Examination of persistent influences evident in Australian contemporary percussion practices throughout the years between 1960 and 2010 takes place in this chapter, by looking back, looking out, looking in and looking forward. These include influences from a variety of music practices from North America, Europe, Asia and Africa, influences from urban Australia, the nation's geographical context, and an exploration of artistic approach and community. Chapter Seven thus develops the meanings of Australian percussion activity between 1960 and 2010 as applied to contemporary music globally, and is vitally engaged with ideas of current contemporary practice and is representative of cutting-edge critical thought in the field. As the first in-depth examination of contemporary percussion activity from an Australian perspective, the aim of this book is to serve as a starting point for further investigation in the field for performers, scholars and educators.

Contemporary percussion: a brief international history

In order to contextualise the forthcoming in-depth documentation of significant events in Australian percussion history, a brief summary of significant developments in early international percussion practices will first be presented. Although percussion instruments have existed for millennia, contemporary percussion is considered a twentieth-century innovation. Until the early twentieth century, percussion instruments were used in a Western art music context primarily for colour, sound effects, humour or novelty value. This began to change in the 1930s, as North American and European composers including Amadeo Roldan, Edgard Varèse, John Cage, Johanna Beyer, Henry Cowell and Lou Harrison began composing repertoire that utilised percussion instruments as concert instruments in their own right. These works and others composed in the 1930s and 1940s marked the beginning of contemporary percussion activity and the subsequent elevation of percussion instruments to an equal status with other instrumental families.

Arguably the most important of these works is Edgard Varèse's *Ionisation*, composed between 1929 and 1931. Written for 13 players, *Ionisation* was not a transcription of music for other instruments, as had been common practice up until this time. *Ionisation* was instead composed directly for the instrumentation in the final ensemble, which included snare drum, temple blocks, gong, siren, guiro, tambourine, triangle and bells. Heralded as the first work composed exclusively for percussion instruments, *Ionisation* is a landmark within the percussion repertoire. At the same time, in 1930, Amadeo Roldan was also composing his *Ritmica No. 5* and *Ritmica No. 6*, and there is some debate surrounding the question of which of these two composers was truly the first.

Varèse and Roldan's groundbreaking works for percussion ensemble were followed by Henry Cowell's *Ostinato Pianissimo* in 1934. Two of Cowell's students, Johanna Beyer and Lou Harrison, also composed for percussion ensemble during the genre's formative years in North America. Beyer's often overlooked percussion output commenced prior to Cowell's, in 1933, with *Percussion Suite in Three Movements*. Beyer composed five more percussion works in the 1930s: *IV* (1935), *March for 30 Percussion Instruments* (1939), *Percussion*, opus 14 (1939), *Three Movements for Percussion* (1939) and *Waltz for Percussion* (1939). Her contemporary, Lou Harrison, produced a number of ensemble works featuring percussion instruments, including *Bomba* (1939), *Concerto No. I for Flute and Percussion* (1939), *Concerto for Violin and Percussion Orchestra* (1940), *Canticle No. 1* (1940), *Canticle No.* 3 (1940), *Labyrinth* (1941), *Fugue* (1941), and *Suite for Percussion* (1942). Harrison regularly collaborated with John Cage on percussion concerts in the 1930s and 1940s, and the two were responsible for pioneering the use of found objects as percussion instruments. An example of this can be found in their co-composed work *Double Music* (1941).

John Cage is celebrated amongst percussionists for his contributions to early percussion practices and percussion ensemble repertoire. Cage wrote a number of works for percussion ensemble in the 1930s and 1940s, and in the 1940s, formed ensembles and taught instrumentalists how to play his percussion works, including *First Construction in Metal* (1939), *Second Construction* (1940), *Third Construction* (1941) and *Amores* (1943). Cage's ensembles included friends of the composer who were dancers, actors or musicians. The fact that untrained percussionists originally performed Cage's early percussion works was not unusual, as there were few opportunities for formal percussion training at the time. What is interesting, however, is that it suggests that composers rather than percussionists motivated the rise of the percussion ensemble in North America (Schick, 2006a: 66). Some decades later, the influence of composers was still present in the formation of international professional percussion ensembles, and later still in professional Australian percussion ensembles, and this will be explored in the forthcoming chapters.

The suggestion that composers led the initial drive to create works for percussion perhaps explains the lack of solo percussion repertoire created during the same period. As Schick outlines, many of Cage, Beyer, Harrison and Cowell's percussion works of the 1930s and 1940s were devoid of any technically advanced material – such as snare drum rolls or rudimental decorations, which would have proved difficult for an untrained performer to execute – but were possible for other instrumentalists to achieve on percussion instruments (Schick, 2006a: 66). Without the range of compositional techniques appropriate to ensemble composition, a reasonable portion of the success of any solo percussion repertoire would have relied on the performance skills of a trained or practised percussionist. A complete list of significant percussion ensemble works composed prior to 1970 can be found in John H. Beck's *Encyclopaedia of Percussion* article 'Percussion Ensemble' (J. H. Beck, 1995: 272). Although percussion ensemble works gradually began to appear more frequently in North America and Europe over the next 30 years, a delay occurred with regard to solo multiple percussion, performances of which would not emerge until the late 1950s. American percussionist Allen Otte suggests that this delay was due to the change in direction of composers who were interested in exploring percussion music. He proposes that those composers who initially investigated the possibilities of percussion found 'mediums other than percussion more useful in realizing their desires, and the need for a [solo] literature to follow that of the '30s and '40s was, at least momentarily, unfulfilled' (Otte, 2013).

In the late 1950s, a small number of individuals began performing solo multiple percussion music. Composers such as Karlheinz Stockhausen (1928–2007), Josef Anton Riedl (1929–2016), Roman Haubenstock-Ramati (1919–1994) and Sylvano Bussotti (b. 1931) began composing solo percussion works at this time, followed by James Tenney (1934–2006) and Morton Feldman (1926–1987) in the 1960s and Iannis Xenakis (1922–2001) and others in the 1970s. Performed by Ray des Roches, Jan Williams and Max Neuhaus in North America, in Germany by Siegfried Rockstroh and Christoph Caskel, and in France by Jean-Pierre Drouet and Sylvio Gualda, this music became the foundation of a solo percussion canon. Table 0.1 shows a complete list of solo multiple percussion works composed between 1950 and 1970.

Composers who were searching for new ways to write music continued to drive the international repertoire for percussion solo and percussion ensemble music. They were joined by percussionist-composers such as John Bergamo (1940–2013). Otte suggests percussionists like Bergamo became interested in writing their own solo repertoire in the 1960s in an effort to continue to sustain the percussion activity instigated in the previous decades, and to balance the literature created in order to sustain contemporary percussion activity (Otte, 1973).

Although close collaboration between percussionist and composer formed an important part of the development of this repertoire, percussionists were not yet the driving force behind the advancement of international percussion repertoire. Instead, they demonstrated techniques and contributed to the sourcing of new instruments and timbres. Examples of works that benefited from this kind of close collaboration between composer and percussionist include *The King of Denmark* (Feldman, 1964), where the performer is instructed to play the work using their

Table 0.1 Published solo multiple percussion repertoire, 1950–1970[1]

Year	*Composer*	*Work*	*Composed for*
1956	John Cage	*27'10.554" for a percussionist*	No one[2]
1957	Josef Anton Riedl	*Stucke*	Unknown
1959	Karlheinz Stockhausen	*Nr. 9 Zyklus*	Christoph Caskel
1959	Sylvano Bussotti	*Coeur pour Batteur* from *Sette Fogli: una collezione occulta*	Unknown
1963	Josef Anton Riedl	*Geschwindigkeit*	Unknown
1964	Morton Feldman	*King of Denmark*	Max Neuhaus
1964	John Bergamo	*Tanka*	Jan Williams
1965	James Tenney	*Maximusic*	Max Neuhaus
1966	Charles Wuorinen	*Janissary Music*	Raymond des Roches
1966	Helmut Lachenmann	*Intérieur I*	Christoph Caskel
1967	Herbert Brün	*Plot*	Michael W. Ranta
1967	Herbert Brün	*Touch and Go*	G. Allan O'Connor
1967	Herbert Brün	*Stalks and Trees and Drops and Clouds*	William Youhass
1969	Roman Haubenstock-Ramati	*Batterie*	No one[3]

fingertips rather than mallets. The instruction to play with fingertips was a suggestion made by percussionist Max Neuhaus during a workshop session with Feldman as the composer continued to request Neuhaus play more quietly (DeHart, 2010: 25). Another well-known example of close collaboration is in the instrumentation of *Psappha* (Xenakis, 1975). Xenakis' explorations of scrap metal yards with percussionist Sylvio Gualda yielded the inclusion of metals in the instrumentation for this work (Schick, 2006b: 5).

By the early 1970s, small but active contingents of composers and performers had established burgeoning contemporary percussion practices in parts of North America and Europe, led by the activities of various professional percussion ensembles. The first of these ensembles was Les Percussions de Strasbourg, founded in France in 1962. Over the next 12 years, they were joined by Canadian group NEXUS (1971), North American groups New Jersey Percussion Ensemble (1968) and Black Earth Percussion Group (1972) and French group Trio le Cercle (1974). As previously mentioned, a small number of solo percussionists were also active in the performing and recording of new works, and the first solo percussion LP *Electronics and Percussion: Five Realizations by Max Neuhaus* had been released in 1968 (Neuhaus, 1968). This was followed by *Cycle for One Percussionist* (Stockhausen, Caskel, Neuhaus, & Rzewski, 1970) and *Pour la Danse* (Gualda & Drouet, 1975). Accordingly, the efforts made by these artists and others gradually began to create a global awareness of contemporary percussion practices. These practices reached Australian shores in the early 1970s, at a time when significant change was taking place in the Australian cultural landscape. This book examines contemporary percussion activity during the period between 1960 and 2010, with discussion of events taking place at the nexus of contemporary percussion history and Australian music history.

Notes

1 In addition to these published works, one unpublished work by Josef Anton Riedl titled *Stück for Schlagzeug* (1951) has been mentioned by Gustavo Aguilar (Lewis & Aguilar, 2014: p. 15).
2 Composed in 1956, *27'10.554" for a Percussionist* was not premièred until 1962 by Siegfried Rockstroh. Personal communication with a number of Cage specialists, including The John Cage Trust, Allen Otte and Steven Schick, in combination with an examination of the 1962 CF Peters Cage Catalogue (Dunn, 1962) suggests that the work was composed by Cage as part of his time length series of works, without a percussionist or première in mind. That Cage appears to have composed this work unprompted is a fascinating situation that is worthy of a separate research investigation.
3 Personal communication with a number of prominent soloists associated with this work has not revealed definitive proof of who this work may have been composed for. During three separate informal conversations between September and December 2017, it was suggested by Jan Williams, Steven Schick and Robyn Schulkowsky that the work was likely composed for its own sake, without a specific performer in mind.

References

Appleby, R. 2012. *Women of Note: The Rise of Australian Women Composers*. Fremantle: Fremantle Press.

Beck, J. H. 1995. "Percussion Ensembles." In *Encyclopedia of Percussion*, edited by J. H. Beck, 269–273. New York: Garland Publishing Inc.

DeHart, J. 2010. "Tap Routes: The Changing Role of the Contemporary Artist-Percussionist." DMA diss., University of California, San Diego.

Dunn, R. 1962. *CF Peters John Cage Compositions Catalogue*. New York: Henmar.

Esler, R. 2014. "I Drum the Body Electric." In *The Modern Percussion Revolution: Journeys of the Progressive Artist*, edited by K. Lewis and G. Aguilar, 112–132. New York: Routledge.

Feldman, M. 1964. *King of Denmark*. New York: Peters.

Gualda, S., and J-P. Drouet. 1975. *Pour la danse*. LP. France: Unidisc.

Hartenberger, R., ed. 2016. *The Cambridge Companion to Percussion*. Cambridge: Cambridge University Press.

Hooper, M. 2012. *The Music of David Lumsdaine: Kelly Ground to Camberwarra*. London: Routledge.

Kerry, G. 2009. *New Classical Music: Composing Australia*. Sydney: New South Wales University Press.

Lewis, K., and G. Aguilar, eds. 2014. *The Modern Percussion Revolution: Journeys of the Progressive Artist*. New York: Routledge.

McNeill, R. 2014. *The Australian Symphony from Federation to 1960*. Vermont: Ashgate.

Neuhaus, M. 1968. *Electronics and Percussion: Five Realisations by Max Neuhaus*. Columbia Masterworks. Compact disc.

Otte, A. 1973. "Preferences in Percussion." Michigan State University. www.msu.edu/~sullivan/OttePref.html

Priest, G. 2009. *Experimental Music: Audio Explorations in Australia*. Sydney: University of New South Wales Press.

Rogers, V. 2009. *The Music of Peggy Glanville-Hicks*. Farnham: Ashgate.

Schick, S. 2006a. *The Percussionist's Art: Same Bed, Different Dreams*, 1st ed. New York: University of Rochester Press.

Schick, S. 2006b. "X Is for Xenakis." [liner notes]. In *Xenakis: Percussion Works*. Mode 171/73. Compact disc.
Schick, S. 2017. "Keynote Presentation." Paper presented at the Transplanted Roots Percussion Research Symposium, Brisbane, 23 September.
Stockhausen, K., C. Caskel, M. Neuhaus, and F. Rzewski. 1970. *Cycle for One Percussionist*. Montreal, Canada: Polydor Records.
Symons, D. 2016. *Before and After Corroboree: The Music of John Antill*. Farnham: Ashgate.
Xenakis, I. 1975. *Psappha*. Paris: Editions Salabert.

Personal communication

Otte, A. 2013. Email Communication with Louise Devenish, 2 July.
Tomlinson, V. 2017. Personal Communication with Louise Devenish, 10 November.
Ughetti, E. 2017. Personal Communication with Louise Devenish, 4 December. Melbourne: Victoria.

1 Contemporary percussion in Australia

The 1960s and 1970s

Prior to 1970 there was very limited awareness in Australia of the contemporary percussion practices that had been emerging in North America and Europe since the 1930s.[1] Australia's geographical isolation was a contributing factor in the delayed development of contemporary percussion music there, simply because the innovations of European and American percussionists and composers had not yet reached Australian shores. This was also true for other forms of contemporary music, including experimental, avant-garde, electronic and improvisatory new music, which began to take root in Australia in the 1960s. The years between 1960 and 1975 saw rapid changes in the Australian cultural climate, changes that made room for the new practices of contemporary percussion music. Until the 1960s, many Australian composers and performers had identified largely with the English pastoral style of creating music and focussed largely on classical and neoclassical forms, genres that traditionally do not focus on percussion instruments. Described as 'an unparalleled and unique sociological and cultural revolution' (Murdoch, 1975: xi), Australian musical activity during this period reflects 'a new-found self-confidence' (Rogers, 2009: 242) amongst musicians. Increased recognition via the presentation of new professional and educational opportunities in Australia was inspiring a new generation of Australian composers who embraced influential modernist musical styles. Greater awareness of and some level of pride in Australian contemporary music making followed, and this was recognised at a federal level in the late 1960s.

During his time as Prime Minister (1968–1971), John Gorton's government established a number of federally funded arts boards designed to support film, literature, music and the visual arts: the Commonwealth Art Advisory Board, Commonwealth Literary Fund and Advisory Board, Australian National Film Board and Commonwealth Assistance to Australian Composers (CAAC) Advisory Board. Composed of Bernard Heinze, Frank Callaway, John Hopkins and later Ernest Llewellyn, CAAC provided financial assistance in the form of fellowships and grants, supporting the creation of new Australian compositions. Following his election as Prime Minister in 1972, Gough Whitlam's government amalgamated these boards to form the Australian Council for the Arts, later the Australia Council. The amalgamation was accompanied by a significant increase in funding capacity over three years, from 7 million dollars to 24 million dollars (Whitlam,

1985: 563). This led to the establishment of many small performing arts groups, and artists of all disciplines began taking greater creative risks in their work, resulting in greater diversity of output. In 1975, the Australian Music Centre was established to document and support the work of Australian composers.

Additionally, the return of expatriates and immigration of international musicians aided this change. In reflecting on this, Warren Burt refers to the beginnings of a contemporary music climate in 1960s and 1970s Australia when he states that 'much of the energy, and many of the new developments in Australian music seem to be the result of migrant musicians, or of Australians returning home after extended periods overseas' (Burt, 2007). Although Burt refers primarily to the contributions made by experimental and avant-garde composers, this proves to be true for the development of Australian contemporary percussion as well. The efforts of the composers to whom Burt refers – for example, Keith Humble and Felix Werder in Melbourne and David Ahern in Sydney – contributed to the development of Australian percussion music and ought to be mentioned.

Australian composer Keith Humble (1927–1995) returned to Melbourne for the second time in 1966 after an extended period of working in electronic and experimental music in Paris, where he had directed the world-renowned performance space Le Centre de Musique. Humble founded Le Centre de Musique at the American Artists Centre in Paris in 1959 and Le Centre operated until 1968. In addition to bringing numerous ideas about experimental music with him to his students at the Melbourne University Conservatorium of Music, Humble invited his international colleagues to engage with Australian musicians within the department. One of these colleagues was French percussionist and composer Jean-Charles François, who, in coming to Melbourne in 1969 at Humble's invitation, was highly influential as a composer. At this time, François was the only percussionist living in Australia with any experience of contemporary percussion performance practices. While François' arrival presented the first opportunity for an introduction of contemporary percussion practices to Australia, local musicians and audiences were initially not receptive to this style of music and most of François' contributions within the Australian contemporary music climate of the time were made within the field of experimental composition. In an interview with John Whiteoak, Humble reflected that during this period at Melbourne University, François' work ethic demonstrated to composers there – who were very enthusiastic about contemporary composition, yet lacking in relevant musical or performance skills – one way to reach the artistic standards they desired:

> I was lucky in the sense that I had a young colleague, Jean-Charles François in France who had a more radical platform than myself in relation to the composition/performance of contemporary music. He particularly wanted to go to the United States, but I said: 'Don't go to the United States straight away, come to Australia?' So he did come to Australia. He had a platform which was basically: it really didn't matter how much talent you had, it was a question of getting the work done. He persisted, and he worked very, very hard while he was here. He also had to realise that there was, culturally, a

> difference between that French idea of: it didn't matter if you were professional or not, you learnt mechanically to play an instrument, and you learnt in life to be efficient, whereas in Australia we're not very efficient. The process of discovering this hurt. So, to answer your question: yes, it was a particular problem.
>
> (Whiteoak, 1989)

As a percussionist, François began to work on a casual basis with the Melbourne Symphony Orchestra, and in 1970 he accepted the position of principal timpanist. Following this appointment, a position teaching orchestral percussion was created for him at the Melbourne University Conservatorium of Music. Despite his regular orchestral work, François' main interest remained in contemporary music, which he attributes to the French musical climate in which he trained (François, 2012). François reflects that in the late 1960s, many Australians had 'a strong opposition to any forms of contemporary arts' and much of the activities at Melbourne University under Keith Humble's direction were simply efforts to introduce new repertoire and contemporary music practices such as electronic music and improvisation to Australian audiences (François, 2012). Occasionally, François performed solo percussion concerts that included his own compositions, but he found:

> Critics in Australian newspapers were very negative towards my activities, they could not accept the Dadaist aspects of many of my performances and compositions. But when I played a farewell recital in Melbourne before my return to France, the Herald critic who had been the most negative before, titled his article 'Farewell to One of the Best'.[2]
>
> (François, 2012)

A small group of experimental musicians, many of whom were based at the Melbourne University Conservatorium of Music, shared François' passion for contemporary music. François had some influence on a small number of young percussionists during his time in Melbourne, in that some awareness of contemporary percussion practices was gleaned. One such student was John Seal, whose impact on contemporary percussion in Melbourne will be discussed further in Chapter Two. However, the lack of interest or skills to perform contemporary percussion music from both his students and François' professional orchestral percussionist colleagues, coupled with the relatively short period he was in Melbourne, prevented François from having a strong or long-term impact on percussion in Australia. Likewise, Australia's isolation proved both productive and frustrating for François. A lack of performance opportunities for a contemporary percussionist in Melbourne at that time allowed François to thoroughly practice the big works of the solo multiple percussion repertoire of the day – Stockhausen's *Nr. 9 Zyklus* (1959), Feldman's *The King of Denmark* (1964) and Wuorinen's *Janissary Music* (1966) – however without the opportunity to perform them. Consequently, François left Melbourne in 1972 in order to pursue contemporary music in a more vibrant environment. Shortly after his return to France he was appointed Head of Percussion at the University of California San Diego, whose music department was renowned for its progressive methods at that time. François also published a number of articles on new music, including 'Organization of Scattered Timbral Qualities: A Look at Edgard Varèse's Ionisation' (François, 1991), which has become one of the most respected analyses of the work. Although these contributions were made following his departure from Australia, François believes

> without that stay [in Australia] far away from the rest of the world, my career in America would not have been as successful. It is difficult for me to measure in return what impact I had on the Australian musical community and percussion playing.
>
> (François, 2012)

Similarly, Humble left Australia around the same time. Although Humble wanted to live and work in Australia, the lack of opportunities for contemporary music in Australia in the late 1960s forced him to leave Melbourne in 1971 to take up a teaching post at the University of California San Diego. Humble returned to Melbourne in 1974 when an opportunity to establish a contemporary music department arose at La Trobe University, where he was appointed Foundation Professor. By this time, contemporary music awareness was beginning to grow and Humble remained in this post until his retirement in 1989.

Professional training for Australian percussionists

The first pocket of Australian percussion activity outside the orchestra arose in South Australia in the 1960s. Richard 'Dick' Smith (1911–1991), principal timpanist with the Adelaide Symphony Orchestra (ASO) was at the centre of this burgeoning percussion scene as a performer and educator. A largely self-taught musician, Smith was passionate about percussion and sound. As a young man, he performed as a xylophone and traps percussionist in pit orchestras for silent cinema. His interest in 1920s ragtime xylophone music further led him to develop his xylophone technique by ordering xylophone lessons from American virtuoso George Hamilton Green. Originally available by mail order, these lessons have since been published as a collection entitled *Instruction Course for Xylophone* (Eyles & Whaley, 1984) and are lauded as the definitive xylophone method book. Ragtime xylophone music is virtuosic and spectacular by nature and Smith's solo xylophone performances gained him a reputation as the leading percussionist in South Australia.

Smith, shown in Figure 1.1, further pursued his interests in orchestral percussion, percussion ensemble, repertoire and history by forming professional relationships with percussionists, composers and percussion instrument manufacturers. According to former student Ian Cleworth, Smith corresponded with distinguished percussionist and author James Blades, xylophone virtuosi Red Norvo and Joe Green and orchestral percussionists Charles White and Bruno Valter (Cleworth, 2013). Such correspondence ensured Smith and his contemporaries in

Figure 1.1 Richard Smith, c. 1965.

Source: Photograph supplied by Michael Askill.

Adelaide were able to maintain some contact with percussion activity happening elsewhere in the world.

Among Smith's activities at the time was direction of It's Percussion, an ensemble of approximately 12 percussionists who performed arrangements of popular tunes 'in the style of Dick Schory' (Bailey, 2012). Richard 'Dick' Schory directed the Percussion Pops Orchestra (PPO), one of the earliest keyboard-based percussion ensembles in North America. Much of the PPO's music came from Schory's own arrangements of popular and traditional music, and It's Percussion used a similar model. It's Percussion regularly performed live for an Australian Broadcasting Corporation television program of the same name in the 1960s. Figure 1.2 is a photograph of It's Percussion during a recording session in the 1960s, with Smith at the centre of the image.

Figure 1.2 suggests that although It's Percussion was the first instance of percussion being treated as an instrument in its own right in Australia, the ensemble was

Figure 1.2 It's Percussion.

Source: Photograph supplied by Michael Askill.

still firmly identifying with orchestral traditions. The instruments are arranged in a formation similar to that of a symphony orchestra, and the players wear tuxedos as would their orchestral counterparts. Additionally, It's Percussion performed transcriptions of orchestral, jazz and dance repertoire, rather than new contemporary compositions. It's Percussion therefore represents a precursor to contemporary percussion ensembles.

The combination of his work with It's Percussion and the ASO made Smith the natural choice to teach percussion students when the Elder Conservatorium in Adelaide introduced percussion to its single study program in 1964. One of Smith's strengths was his professional approach to percussion performance and education, and this was passed on to his students. James Bailey, one of Smith's first students, speaks of this professionalism as 'a feeling of respect to the profession in that what you were participating in was not a game, it was important and serious' (Bailey, 2012). Bailey describes the format of his percussion training as a single study option:

> At that time there was no such thing as a degree in percussion. It didn't exist in the whole of Australia. Single study was where it was at, so you basically did an audition, got accepted and then you started having a weekly lesson with [Smith] and then he would ask you to participate in a variety of brass ensembles and various activities around the Conservatorium that needed percussion.
>
> (Bailey, 2012)

In 1968, the Elder Conservatorium became the first Australian institution to include percussion in their Bachelor of Music degree. Following the establishment of formal tertiary training at the Elder Conservatorium, a percussion community began to form around Smith's activities. This grew over the coming decades and began to have an effect on percussion in other states of Australia as students of Smith gained various positions around the country. Smith's activities at the Elder Conservatorium set the scene for the contemporary percussion music that was to emerge in the 1970s, and many of his students became leaders in the field in the decades that followed. Table 1.1 shows a timeline pertaining to the establishment of Australian tertiary percussion programs.

Smith's role in the growth of Australian percussion is acknowledged by many percussionists, including Colin Piper, who states: 'In the '60s and '70s, Richard Smith was by far the greatest percussion teacher in the country. I don't think there's any doubt about that' (Piper, 2013). Further, his contributions to Australian percussion practices were recognised at the first Australian Percussion Gathering in 2010 with the posthumous awarding of the inaugural Lifetime Achievement Award. Although Smith's passion and skills lay primarily in orchestral percussion and his teaching was focused on the development of a strong orchestral technique as a foundation,

Table 1.1 Tertiary percussion programs

Institution	*Course established*	*First teacher/s*
Elder Conservatorium Founded 1898	1964 Single Study 1968 Bachelor of Music	Richard Smith
Melbourne University Conservatorium of Music Founded 1901[3]	1969 Single Study[4]	Jean-Charles François
Victorian College of the Arts Founded 1974	1974 Bachelor of Music	Barry Quinn
Sydney Conservatorium Founded 1916	1981 Bachelor of Music	Jack Purdon
Canberra School of Music, Australian National University Founded 1965	1982 Single Study 1985 Bachelor of Music	Graeme Leak Daryl Pratt
West Australian Academy of Performing Arts Founded 1980	1984 Single Study 1987 Bachelor of Music	David Pye Gary France
University of Western Australia School of Music Founded 1957	1986 Bachelor of Music	David Pye
Flinders Street School of Music (TAFE) Founded 1970	1986 Certificate of Music Performance	Ryszard Pusz
Queensland Conservatorium, Griffith University Founded 1957	c. 1975 Single Study 1980 Bachelor of Music	Unknown[5] Richard Mills
Tasmanian Conservatorium of Music Founded 1964	c. 1975[6] Bachelor of Music	Graham Johns
Australian National Academy of Music Founded 1994	2013 Professional Performance Program	Peter Neville

students interested in pursuing other techniques or forms of percussion music not in Smith's repertoire, such as a four-mallet marimba technique, were not discouraged.

Despite a firm connection to orchestral percussion, the percussion studio Smith established at the Elder Conservatorium provided, for the first time in Australia, a base for percussion students to explore various forms of percussion music at a tertiary level and, as a result, a way into the profession. A pattern began to emerge as Smith's students frequently found work in orchestras, hotels and big bands around Australia and in Asia on Smith's recommendation. These appointments gave emerging professionals the opportunity to continue to develop their professional skills. When those percussionists who had travelled overseas returned to Australia, they brought with them new skills and ideas to which they had been exposed. For example, Bailey followed this path and spent nine years working in Hong Kong as well as studying drum set and jazz vibraphone in America. Upon his return in 1976, Bailey introduced another contemporary technique to students in Adelaide: the Gary Burton four-mallet vibraphone technique. The emergence of four-mallet keyboard technique in North America and Europe elevated the marimba and vibraphone to a status on a par with other concert instruments more quickly than other percussion instruments. It represented a virtuosic technical aspect of percussion performance and enabled composers to write more complex material for performers. As a result, four-mallet practices became common elsewhere, however it was virtually unheard of in Australia at this time, where classical and jazz mallet players still favoured a two-mallet approach to keyboard percussion instruments. Bailey reflects that there was a certain amount of resistance to learning four-mallet technique in the beginning:

> It was still a two-mallet approach, whether it was jazz or classical players. There was a certain amount of resistance [to learning four-mallets] because I think people didn't fully appreciate the concept of what four-mallet playing was about. It was a process of re-educating.
>
> (Bailey, 2012)

Coupled with the changes in Australian music making outlined previously, the presence of Smith's tertiary percussion program and the successful entry of his students into the profession were signs that the musical landscape in Australia was preparing for the introduction of contemporary percussion activity.

Les Percussions de Strasbourg

The catalyst for true contemporary percussion activity in Australia was the tour of the French ensemble Les Percussions de Strasbourg in 1971. Founded in 1962 at the suggestion of composer and conductor Pierre Boulez, Gabriel Bouchet, Jean Batigne, Georges Van Gucht, Claude Ricou, Jean-Paul Finkbeiner and Détlef Kieffer were the founding members of the world's first professional percussion ensemble. Still active today, this ensemble has contributed much to the global field of contemporary percussion via commissioning and premièring hundreds of works for percussion ensemble, including many of the most significant works in the repertory. From the outset, the group's impact on how composers, colleagues and the general public viewed percussion was profound, and they have been described as 'pioneers in their field. . . . Their commitment to new compositions inspired composers and audiences alike with a sense of urgency in the quest for innovative forms of musical expression' (Bernard, 2011).

Les Percussions de Strasbourg's work in the field of contemporary percussion was recognised internationally and came to the attention of British composer, conductor and new music advocate John Hopkins (1927–2013) in the 1970s. In 1963, after a period of six years in New Zealand, Hopkins migrated to Sydney to take up the position of Federal Director of Music at the Australian Broadcasting Commission. Actively engaged in the promotion of new music, Hopkins frequently programmed new works by emerging and established Australian and international composers in ABC broadcasts and the concerts he conducted. Hopkins became an inspiration for young musicians as conductor of the Australian Youth Orchestra and later as the founding Dean of Music at the Victorian College of the Arts. In light of this, Hopkins can be considered a key figure in the development of Australian percussion, as it was he who invited Les Percussions de Strasbourg to tour Australia after meeting them in Europe.

Supported by the ABC, the Australia Council and classical music promoter Musica Viva, Les Percussions de Strasbourg performed repertoire especially commissioned for their visit with ABC symphony orchestras and concerts of their own percussion ensemble repertoire around the country in 1971. Their performances in Perth, Adelaide, Melbourne, Sydney and Canberra showcased large-scale virtuosic percussion repertoire. Indeed, this was the first time contemporary percussion music of this scale had been performed in Australia and it was a revelation for artists around the country.

Already an influence on artists in North America (DeHart, 2010: 26), Les Percussions de Strasbourg drew specialist Australian audiences who were exposed for the first time to the European contemporary percussion culture. Similarly, many Australian percussionists cite Les Percussions de Strasbourg performances as significant moments in their musical development. In a 1994 *Sydney Morning Herald* article, for example, Colin Piper states: 'They were the first *real* percussion ensemble and I was quite bowled over by the music. It opened my eyes and ears to the possibilities that existed in percussion' (Adamson, 1994: 61). Likewise, James Bailey describes Les Percussions de Strasbourg as 'the first real, serious, high end contemporary group that came [to Australia] and they re-orientated percussion thinking to something that was a totally different concept to what had existed previously in Adelaide' (Bailey, 2012). Also in Adelaide was a young percussionist named Michael Askill, who had become familiar with Les Percussions de Strasbourg by collecting their LP recordings. A family subscription to the Australian World Record Club fostered Askill's interest in world music, contemporary music and jazz, enabling the teenage Askill to discover contemporary music by Messiaen, Boulez and Les Percussions de Strasbourg. Operating between 1957 and 1976, the Australian World Record Club was a franchise of the UK-based mail order company through which members could purchase recordings of jazz and classical music. Askill believes the Australian World Record Club was highly influential on Australian musicians in the 1960s and 1970s because it enabled them to purchase international records of all genres of music that were not necessarily available in local stores and therefore keep abreast of some international trends (Askill, 2013, 2015). Listening to Les Percussions de Strasbourg recordings was beneficial for the young Askill; however, experiencing their live performance program cemented the ensemble in his memory. In a 2013 interview, Askill remembered: ‘I had never seen or heard a group play their kind of repertoire so well and it was pretty awe-inspiring’ (Askill, 2013). Correspondingly, Tim White reflects on the 1976 Canberra concert thus:

> It was like the doors in Australia had finally been flung open to this world of new composers, new sounds, new ways of doing things and it was done really seriously by very talented players. You’ll see references by non-percussionists too whose lives were changed by Les Percussions de Strasbourg, it was something that they’d never conceived possible.
>
> (White, 2012)

There were four primary reasons Les Percussions de Strasbourg had such a strong impact on so many. First, their tour encompassed five of the eight Australian capital cities, including Perth, Adelaide, Melbourne, Sydney and Canberra. A national tour of this scale was unprecedented for a contemporary music ensemble. Second, they performed repertoire especially commissioned for their tour with ABC symphony orchestras, and their existing repertoire in chamber music concerts. Their decision to engage with Australian musicians by performing Australian works with Australian orchestras served to recognise and engage with the local music scene. Third, their existing ensemble repertoire included the entire percussion ensemble catalogue to date. Finally, this repertoire was large-scale and virtuosic, requiring large and complex set-ups of unusual instruments.

Les Percussions de Strasbourg’s repertoire consisted of works composed for, or arranged by, members of the group. This repertoire highlights the ensemble’s connections with leading figures in contemporary music, including Edgard Varèse, who had previously ‘spontaneously agreed’ to give Les Percussions de Strasbourg permission to arrange a reduced version of his landmark work *Ionisation* from 13 players to six.[7] The concert program presented at the University of Western Australia on 25 September 1971 and the extensive instrument list that accompanied this program is shown in Figure 1.3 and Table 1.2.

Table 1.2 shows an astonishing instrument list. Apart from its sheer size, many of the instruments were unusual and could not be sourced in any Australian city in the 1970s. To contextualise this list, it is worth noting that the smaller and more common 4.3 octave marimba was only just beginning to be used in Australia at this time and the five-octave marimba was relatively new even to cities where contemporary percussion was flourishing. Additionally, many of the exotic percussion instruments, including Thai gongs, mokubyos (Japanese woodblocks),

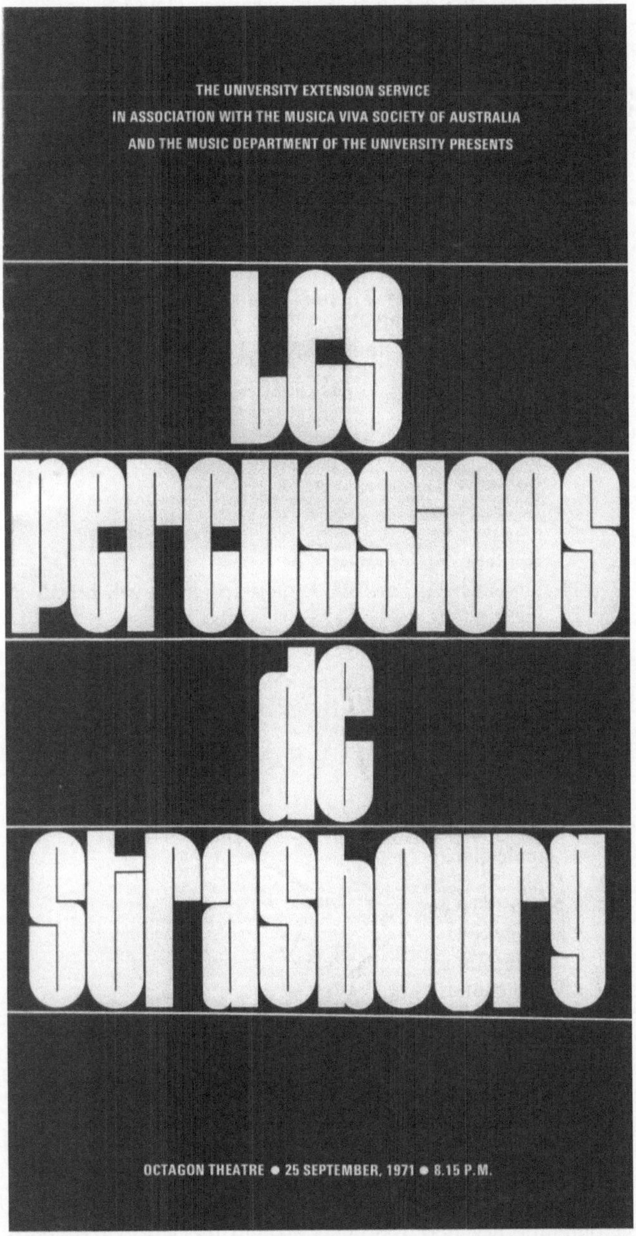

Figure 1.3 Cover of Les Percussions de Strasbourg 1971 program at the University of Western Australia.

Table 1.2 Les Percussions de Strasbourg 1971 program and instrument list

Program	MILOSLAV KABALÁC *Huit Inventions* Opus 65 GILBERT AMY *Cycle* EDGARD VARÉSE *Ionisation* KAZIMIERZ SEROCKI *Continuum*
Instrument list	1 5.0 octave Marimba 1 Xylorimba 2 Vibraphone 2 Hand tuned or single pitch timpani + 2 pedal timpani 2 Pedal timpani 2 Timbales créole 13 Turkish suspended cymbals (between 26 cm and 72 cm) 2 Chinese cymbals + 2 normal cymbals 2 Pairs of crash cymbals 6 Tam-tams: 1 very small, 1 small, 2 medium, 1 large, 1 very large 2 Hi hats 12 Thai gongs 4 Triangles 2 Chromatic sets of crotales 1 Set of bells 2 Glockenspiels 5 Cowbells 3 Sets of chromatic cowbells 11 Tam-tams: 3 groups of 3 (small-medium-large), 1 medium and 1 large 2 Congas 2 Pairs of bongos 3 Large snare drums 1 Snare drum 1 Piccolo snare drum 5 Tambourines 2 Kick drums 2 Concert bass drums 4 Pairs of maracas (2 large, 1 medium, 1 small) 9 Temple blocks 2 Whips 5 Pairs of claves 6 Mokubyos 12 Tablas-Tarang 2 Wood chimes

chromatic crotales, chromatic cowbells and tam-tams, were not only unable to be sourced in Australia but were unfamiliar to Australian percussionists.

An instrument list of this size for a touring percussion ensemble is large even by today's standards. The extensive costs of touring with such heavy and bulky instruments have always presented a significant financial hurdle for percussion ensembles. Contemporary percussionists around the world frequently commission and compose works that specifically reduce the number and size of instruments required in an effort to reduce the costs of touring with instruments. One of the most famous examples of this is Brian Ferneyhough's *Bone Alphabet* (1991). In the program notes that accompany the score, Ferneyhough states that '*Bone*

Alphabet came about as the result of a request by Steven Schick for a solo work for a group of instruments small enough to be transportable as part of the performer's personal luggage when travelling by air. I responded by leaving the precise instruments to be utilised unspecified' (Ferneyhough, 1991). Similarly, in a 2012 commissioning project, the Australian duo The Sound Collectors gave composers a list of instruments they were prepared to travel with and a description of their vehicles to give an indication of the amount of space they would have for instruments. Like Ferneyhough, all four of the composers involved opted to leave instrument choice up to the performers.

That funding was found to freight Les Percussions de Strasbourg instruments to Australia can be attributed to two factors: first, the quality of the group's work, and, second, the group's desire to remain faithful to the composers whose works they presented. It is a quirk of multiple percussion music that no two performers' setups of a solo work or ensemble part will be the same due to the instrumentation available to each performer. A composer's request for a percussionist to select a suspended cymbal and a cowbell can result in hundreds of different sonic possibilities dependent on the instruments available to the said percussionist. The seemingly simple directive of 'suspended cymbal' can refer to a cymbal of any diameter between 8 and 24 inches and could lead a performer to consider numerous options regarding timbre – ride, crash, splash or suspended orchestral cymbal – and brand. The probability of a percussionist in Sydney selecting the same size, brand and type of cymbal as a percussionist in Strasbourg is extremely low, due to personal preferences and each percussionist's instrument collection. Even in the unlikely event that two players would have a matched instrument collection, the nature of cymbal making would result in a variation of pitch and sound quality, well before consideration of mallet choices comes into the picture.

With this in mind, it is likely that the preference by Les Percussions de Strasbourg percussionists to tour with their own instruments, coupled with the high professional quality of their work, were reasons enough to bring their instruments to Australia. This decision was taken on their 1971 and 1976 tours; and the subsequent presentation of dozens of unusual percussion instruments to Australian players was one element of their performances that profoundly affected Australian percussionists. As a member of the audience for the group's 1976 Canberra concert, percussionist Tim White states:

> I've got a vivid memory still of the lions roar in *Ionisation*, and I think they played *Ionisation* with six players and the guy was playing from above, it was suspended from above and [he] was pulling down on the rope. I thought this was superb . . . here were these guys [sic], they looked like magicians to me. (White, 2012)

Additionally, Les Percussions de Strasbourg's 1971 tour provided an impetus for two Australian composers to write serious contemporary percussion ensemble music for virtuosic players, the first time any Australian composers had considered writing for this type of ensemble. Les Percussions de Strasbourg commissioned the first Australian percussion ensemble works: Peter Sculthorpe's sextet *How the Stars Were Made* (1971) and Barry Conyngham's *Six* (1971), for percussion sextet with orchestra. Premièred during their tour alongside established repertoire from Europe and North America, these works were later arranged for performance by Australian percussion ensembles.

The stirrings of percussion activity in 1960s Adelaide were antecedent to the forthcoming contemporary percussion ensembles of the 1970s. It's Percussion and Richard Smith foreshadowed a new perspective on the role of percussion in Australian music activity at a time when wider Australian music practices were commencing a period of significant development. When viewed in light of the Australian cultural climate in the early 1970s, Les Percussions de Strasbourg's 1971 tour provided a timely inspiration for Australian percussionists. A noticeable increase in local activity and interest in exploring contemporary percussion practices followed the first Les Percussions de Strasbourg tour. Inspired by the work of their international peers and keen to collaborate with composers, percussionists began to consider creating percussion performances in addition to their orchestral commitments. A prolific new era of percussion music making followed, spearheaded by the Australian Percussion Ensemble. This was the first professional contemporary percussion ensemble in Australia, and their collaborative creative presence was the impetus for the development of a body of new percussion repertoire. The activities of the Australian Percussion Ensemble and their impact on Australian contemporary percussion development are examined in detail in Chapter Two.

Notes

1 A number of Percy Grainger's (1882–1961) orchestral and chamber works included percussion parts that contained ideas aligned with international contemporary ideas, such as using unusual sound sources (e.g. tuned glasses) and utilising instruments for melody rather than for effect. However, in this study, Grainger's works are not considered to be the first instance of contemporary percussion in Australia. The reason for his exclusion is threefold. First, percussion ensemble and solo percussion works are considered the core of contemporary percussion development, and Grainger did not compose any works specifically for percussion instruments alone. Works that at first glance may seem to be for percussion ensemble, such as *Eastern Intermezzo* (1898–99, percussion ensemble version 1950) are in fact arrangements of his own works. Generating many versions of his own works or of traditional music was a common practice of Grainger's. Others, such as *Bahariyale V. Palaniyandi,* are one of a number of arrangements of anonymous and/or traditional works. Second, Grainger did not compose these works while in residence in Australia therefore they did not have any immediate impact in Australia. Finally, Grainger primarily used tuned percussion instruments such as xylophones, nabimbas and marimbas for melodies. His approach to tuned percussion instruments separates Grainger's percussive writing from the sound and rhythm based explorations of his contemporaries. Thus, Grainger's works will not be discussed here.

2 Although no articles titled 'Farewell to One of the Best' have yet been found, it can be assumed that the critic to whom François refers is former *Melbourne Herald* music critic John Sinclair. Neither a trained journalist nor musician, Sinclair's reviews caused great consternation among musicians, generated lawsuits against the newspaper and gained him an international reputation as a harsh critic (Thomas, 1998: 79–90).

3 Although 1901 is shown here as the Melbourne University Conservatorium of Music foundation date, it is necessary to mention that this is the date the Melbourne University established and funded a Conservatorium. However, following his appointment as Ormond Professor of Music at Melbourne University in 1891, G.W.L. Marshall-Hall established the Melbourne University Conservatorium in 1895 (Radic, 1982: 20).
4 During personal communication via email with François, he stated that 'A position of percussion teacher at the Melbourne Conservatorium was created and I was asked to take it on a part time basis. The Grainger Museum was the place where I could teach and practice' (François, 2012).
5 Further information can be found in Peter Roennfeldt's study 'Beyond the Ivory Tower – Higher Education Institutions as Cultural Resource: Case Study of the Queensland Conservatorium of Music' (Roennfeldt, 2011: 159).
6 Similarly, this date is based on correspondence with Thomas O'Kelly, who began teaching at the Tasmanian Conservatorium in 1981.
7 The program note from the 1971 performance in Perth contains this quote. As Varèse died in 1965, it can be assumed that this arrangement was made some time before Les Percussions de Strasbourg's Australian tour.

References

Adamson, J. 1994. "Synergy's One-Night Stand Has Reached Its 20th Anniversary." *Sydney Morning Herald*, 28 July.

Askill, M. 2015. "Reflections on Composition and Consciousness." PhD thesis, Brisbane: The University of Queensland.

Bernard, J-P. 2011. "Adventures, New Adventures." www.percussionsdestrasbourg.com

Burt, W. 2007. "Some Musical and Sociological Aspects of Australian Experimental Music." *Resonate Magazine*, 31 July. www.australianmusiccentre.com.au/article/some-musical-and-sociological-aspects-of-australian-experimental-music

DeHart, J. 2010. "Tap Routes: The Changing Role of the Contemporary Artist-Percussionist." DMA thesis, University of California San Diego.

Eyles, R., and G. Whaley, eds. 1984. *George Hamilton Green's Instruction Course for Xylophone*. Florida: Meredith Music Publications.

Ferneyhough, B. 1991. *Bone Alphabet*. London: Peters.

François, J-C. 1991. "Organization of Scattered Timbral Qualities: A Look at Edgard Varèse's Ionisation." *Perspectives of New Music*, 29 (1): 48–79.

Murdoch, J. 1975. *Australia's Contemporary Composers*. Melbourne: Sun Books.

Radic, T. 1982. *G.W.L. Marshall-Hall, Portrait of a Lost Crusader: An Introduction to the Marshall-Hall Collection of the Grainger Museum, University of Melbourne*. Perth: Dept. of Music, University of Western Australia.

Roennfeldt, P. 2011. "Beyond the Ivory Tower – Higher Education Institutions as Cultural Resource: Case Study of the Queensland Conservatorium of Music." *History Compass*, 9: 231–245.

Rogers, V. 2009. *The Music of Peggy Glanville-Hicks*. Farnham: Ashgate.

Thomas, A. 1998. "Beware of Snakes, Spiders and Sinclair. John Sinclair (1919-1991), Music Critic for the Melbourne Herald: The early years." *Context: Journal of Music* Research, no 15/16: 79–90.

Whiteoak, J. 1989. "Interview with Keith Humble." *NMA*, 7: 21–26. www.rainerlinz.net/NMA/repr/Humble_interview.html

Whitlam, G. 1985. *The Whitlam Government*. Melbourne: Penguin Books.

Personal communication

Askill, M. 2013. Skype Interview with Louise Devenish, 7 February. Mullimbimby, New South Wales.

Bailey, J. 2012. Skype Interview with Louise Devenish, 10 November. Penang, Malaysia.

Cleworth, I. 2013. Skype Interview with Louise Devenish, 25 March. Sydney, New South Wales.

François, J-C. 2012. Email Interview with Louise Devenish. Paris, France.

Piper, C. 2013. Live Interview with Louise Devenish, 26 April. Sydney, New South Wales.

White, T. 2012. Live Interview with Louise Devenish, 9 October. Perth, Western Australia.

2 The emergence of Australian contemporary percussion

1970s Melbourne

This chapter commences the documentation and analysis of Australian contemporary percussion activity in the context of major international developments in the genre and a changing local cultural climate. Contemporary percussion began to appear in Australian music activity in the early 1970s, as Australian symphony orchestra percussionists began to consider performance opportunities outside their orchestral commitments. Three professional contemporary percussion ensembles in Australia were formed in the 1970s: the Australian Percussion Ensemble in Melbourne in 1972, The Sydney Percussions in 1974 and Adelaide Percussions in 1978. The efforts made by these groups in the 1970s to produce percussion concerts, repertoire and recordings were inspiring to fellow percussionists, composers and students. Consequently, percussion ensembles gradually became the foundation upon which early professional contemporary percussion practices were cultivated. This chapter explores the emergence of contemporary percussion in 1970s Melbourne, with a particular focus on the first professional percussion ensemble established in Australia: the Australian Percussion Ensemble. The contributions of this vital yet largely forgotten ensemble are documented and examined here for the first time, offering an insight into the changing Australian musical landscape during this period.

The Australian Percussion Ensemble (1972–1977)

Although the Australian Percussion Ensemble (APE) was the first professional percussion ensemble formed in Australia, little is known of their activities in the wider percussion community. There is scant documentation about the group and existing documents based on memories contain conflicting information and inaccuracies. Former APE members Wendy Couch and Dalmazio Babaré hold the most comprehensive collections of APE concert programs and newspaper articles, and both Couch and Babaré's archives confirm that the APE's first performance took place in September 1972, after Jean-Charles François left Melbourne and one year after Les Percussions de Strasbourg's first tour to Australia in 1971. These archives also show that the APE was active as a group until 1976. However, individuals from the APE continued to perform together for some months beyond this date, as works such as Warren Burt's *Salon Music* (Burt, 1977) continued to be composed for combinations of APE performers. From this evidence, it can be concluded that the Australian Percussion Ensemble was active between 1972 and 1977.

The APE was founded by Melbourne Symphony Orchestra (MSO) percussionist John Seal (b. 1947). Prior to his appointment with the MSO, Seal had studied percussion privately with Bill Fleming (a contemporary of Richard Smith), and also pursued private contemporary percussion study with Jean-Charles François during his time in Melbourne. At this time, François was the only percussionist in the country who had received any formal training in contemporary percussion and therefore could share first-hand some experience with young Australian percussionists like Seal. François had studied percussion at the Conservatoire National Supérieur de Musique et de Danse de Paris in the 1960s, a time when the city was establishing its reputation as a 'hotbed' for contemporary music activity. When reflecting on this period, François notes that although contemporary percussion training in conservatoria was relatively new in France at the time, it was more abundant than in other parts of Europe. Thus, as a young graduate, François availed himself of the numerous performance opportunities offered to him both in France and England before moving to Australia (François, 2012). With François as both a colleague in the MSO and teacher, Seal had gained some awareness of French contemporary percussion practices prior to the Les Percussion de Strasbourg tour, and formation of the APE was a logical step forward for Seal following their visit to Melbourne.

The APE's six-player format was inspired by Les Percussion de Strasbourg's sextet, and their early concerts included European and North American repertoire performed by Les Percussion de Strasbourg such as Carlos Chávez's *Toccata* (1942) and John Cage's *First Construction in Metal* (1939) and *Third Construction* (1941). The APE's early members included Seal together with his students Ross Munroe, Steven Whately and Wendy Couch with Dennis Johnson and Ormsby Wilkins. As repertoire required and as membership changed, percussionists John Wise, Brenton Muoy, Dalmazio Babaré, Graham Morgan, Ron Ellam, Malcolm Leek, Kevin Makin and New Zealander Paul Sarcich also performed with the APE.

Although the APE, by virtue of their instrumentation, focussed on new works that included percussion, the group's artistic interests covered a broader spectrum. Seal remembers that the APE concerts were 'a variable feast, with an emphasis on creating a contemporary artistic environment' (Seal, 2012). Seal refers to the APE's focus on programming local experimental music by their peers and colleagues alongside works by established international composers including Luciano Berio and Karlheinz Stockhausen, stating that 'in summation, our driving motive was to the avant-garde and to provide an outlet for new music and a platform for new composers' (Seal, 2013). This statement suggests that the promotion of Australian contemporary repertoire was of great importance to the group, and concert programs outlining the works they performed further supports this claim. As the performance of Australian repertoire was a key focus of APE's mission (Sykes, 1973: 32), as opposed to the performance of purely percussion works, the APE's repertoire included mixed instrumental chamber ensembles, interdisciplinary collaboration and works composed by members of the APE. Figure 2.1 shows a

MULTI MEDIA RECITAL

Something new is coming to our Melbourne artistic horizon!

A multi-media recital combining the individual art forms of POETRY, MUSIC, THEATRE, PAINTING, SCULPTURE, DANCE and FILM.

From this exciting whole it is hoped to evolve a new art form in itself.

Most of the material has been created especially with this idea in mind.

NEW MUSIC, A NEW APPROACH, NEW TECHNIQUES, an evening of stimulating new creative effort.

Following in the mood of the Duncan Academy on the Rue de Seine, Paris, the early days of Wigman, Joss, Stravinsky, Mahler, the Train Bleu, Apollinaire, Hemingway at the Select, the vista is boundless with the inspired new talent on the Australian artistic scene.

At the UNION THEATRE, on September 10th - at 8 p.m. - this exciting theatre and music event will feature the AUSTRALIAN PERCUSSION ENSEMBLE; BARRY PITTARD PLAYERS; ROBERT CLARKE, Trombonist; STEVE MARTIN, Melbourne Symphony Orchestra Double Bass Player; BARRY McKIMM, Melbourne Symphony Orchestra trumpeter, and JOHN SEAL, Melbourne Symphony Orchestra percussionist.

The composers will include - John a'Beckett, who besides his creative talents drives a taxi; Felix Werder, well-known music critic; Ian Bonighton, Melbourne Conservatorium composer; and John Seal, whose percussion compositions have many new approaches; Cacioppo, Berio and Haubenstock - Ramati.

DANCE, SPEECH, MUSIC and DESIGN are incorporated in this thrilling recital with an accent rarely realised in Melbourne art circles previously.

ANY PUBLICITY TO PROMOTE THIS NOTABLE EVENT WILL BE GREATLY APPRECIATED.

Enquiries: BARRY BALMER
94 5887
645 1100

Figure 2.1 Excerpt from an Australian Percussion Ensemble media release.

Source: Supplied by John Seal.

media release for another APE program that incorporates both mixed instrumental repertoire and interdisciplinary collaboration. Their mixed instrumental repertoire featured repertoire by Australian composers Keith Humble, Warren Burt, Ron Nagorcka and Ian Bonighton, and interdisciplinary collaborations included works by Seal, Helen Gifford and Jennifer Mackenzie, designed to bring percussionists together with poets, actors and multi-media in performance.

In 1974, the APE raised the level of their performance production by programming their first concert series, titled Three Creative Recitals. Held at the Union Theatre, this series represented a major step up in the way the ensemble operated. It was the first time the ensemble had curated a major concert series and was advertised accordingly, with professionally designed and printed fliers, shown in Figure 2.2. The repertoire performed across this series included the Australian première of seminal percussion solo *Nr. 9 Zyklus* by Stockhausen, major pieces of the international repertoire and numerous Australian works by David Bedford, James Penberthy, Barry Conyngham and Graham Hair, commissioned by the APE for this series and from previous concerts. Figure 2.3 shows a concert program from the APE's Three Creative Recitals series that includes the première of two commissioned Australian works: Keith Humble's *Prime Riff* (1974) for sextet and tape and Helen Gifford's interdisciplinary work *Images for Christmas* (1973).[1]

The APE commissioned 26 works during the five years they were active, and in some cases had considerable influence on the shape of the works they performed. An example can be found in Robert Irving's *Sums* (1973), the handwritten score of which was accompanied by a three-page letter from Irving of explanatory notes regarding the notation. This detailed letter also outlines some notational and technical difficulties that Irving hoped Seal and his ensemble would find solutions to. Other works, such as the aforementioned *Prime Riff* by Keith Humble were delivered fully notated and complete. That the APE was able to commission such a high number of works within a short period points to a change of pace amongst Australian composers with regard to percussion repertoire. APE collaborator Helen Gifford notes that very few composers were active in Melbourne in the 1960s, on a part- or full-time basis (Appleby, 2012: 57). However, this changed in the 1970s as the number of active Australian composers increased and the presence of ensembles such as the APE provided regular performance opportunities for new work. Composers began forming ongoing collaborative relationships with percussionists through the APE, a practice that had not previously taken place in Australia.

In addition to international repertoire and commissioned works, the APE regularly performed their own original content, including a number of works by Seal for APE events, such as *Melbourne Media* (1973) and *Structures* (1974). Together with a number of Australian percussionists, Seal cites North American and European contemporary musicians as a major influence on his contemporary percussion practice in the 1970s, including his compositions. The *Structures* score shown in Example 2.1 does not contain information regarding tempo, dynamic, instrumentation or the number of performers required. These omissions reference the indeterminate practices of John Cage, and a conscious acknowledgement of Karlheinz Stockhausen. *Structures* was to be interpreted by the performer/s in a manner similar to Stockhausen's landmark percussion solo *Nr. 9 Zyklus*, in that the performer could choose where in the score to start performing. This, in combination with the freedom granted the performer to choose instrumentation, register and tempi, resulted in *Structures* existing in various guises. According to Wendy Couch, one version was performed on a set of wooden box drums designed by Seal, shown in Figure 2.4 (Couch, 2014). Often the interpretation was suited to the venues in which the ensemble was to be perform. Seal describes two versions:

> A performance at the Melbourne International Festival of Organ and Harpsichord at St. Peter's Church in East Melbourne produced a very nice,

Figure 2.2 A double-sided flier advertising a 1974 Australian Percussion Ensemble concert series titled Three Creative Recitals.

Source: Supplied by Wendy Couch.

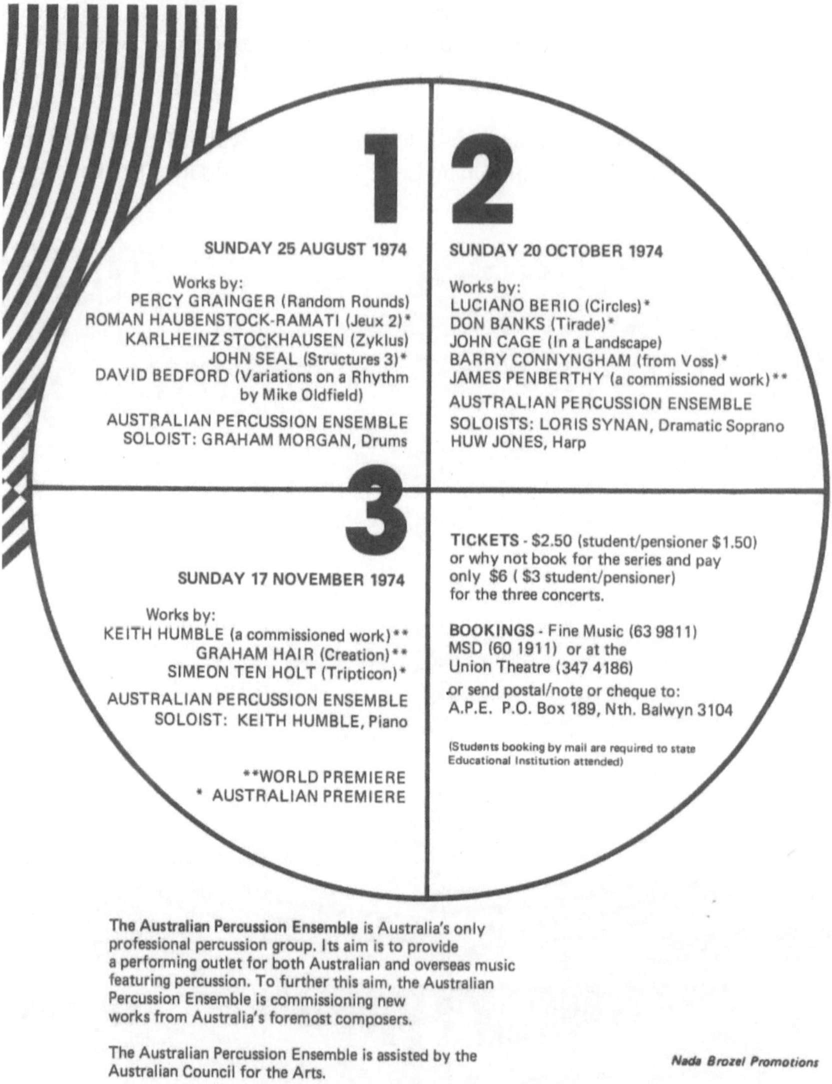

Figure 2.2 (Continued)

meditative interpretation. Another performance of the piece was in a Steve Reich manner of 98:100 as a fugue. For this, we all had to wear headphones with our own dedicated click track. Graham Morgan, one of Australia's greatest ever drum kit players, performed the 100 beat fugue. He started just one beat after the ensemble, in turn finishing each section one beat earlier.

(Seal, 2013)

Figure 2.3 The complete three-page program from the third concert in the Three Creative Recitals concert series.

Source: Supplied by Dalmazio Babaré.

concert three

SUNDAY 17 NOVEMBER 1974
UNION THEATRE 8.15 P.M.

AUSTRALIAN PERCUSSION ENSEMBLE
John Seal, Dal Barbare, Ron Ellam,
Brenton Mouy, Paul Sarcich, John Wise

With Soloists:
KEVIN McBEATH, voice
JOCHEN SCHUBERT, guitar

This concert is presented with the generous assistance of the Australian Council for the Arts.

The Australian Percussion Ensemble

JOHN SEAL, percussionist with the Melbourne Symphony Orchestra, is the founder of this group of six young Melbourne musicians.

Since their formation in September 1972, the Group has appeared in 13 concerts in Melbourne and Sydney, recorded Australian compositions for commercial release, visited schools with lecture-demonstrations and organised workshops for school-children in conjunction with the Grainger Centre in the University of Melbourne. So far the Group has performed 16 new works by Australian composers.

Kevin McBeath

KEVIN McBEATH has been drawn out of retirement as a special favour to composer Helen Gifford to narrate her *'Images For Christmas'* which she has written especially for the Australian Percussion Ensemble.

He has always been interested in promoting contemporary music – both on ABC Radio and at concerts held by the Society of New Music which he ran during the 1950's in association with Alex Berry (now Liaison Officer of the Ministry for the Arts).

Kevin McBeath began his career with the ABC as an announcer. Later he became well-known for his programmes – *At Your Request, New Music, Music Magazine, Byways Of Music,* etc.

He introduced to Australian audiences in 1952 the first live performance of *Facade* in its original setting which resulted in the local recording of this work. Subsequently, the composer chose Kevin McBeath to assist him in his own performances of *Facade* at the 1964 Adelaide Festival.

Kevin McBeath has also been narrator for Honneger's *King David, A Lincoln Portrait* by Aaron Copland and gave the first Australian performance of Schoenberg's *Ode To Napoleon.* These performances led to TV engagements as narrator in *The Soldiers Tale* starring Robert Helpman, and as actor in *Shadow Of Heroes, The Ides Of March, Murder In The Cathedral* and *A Man For All Seasons.*

In 1949 and 1954, he worked with the BBC and toured England and America collecting taped interviews with contemporary composers for his ABC programmes.

A series of throat operations has forced Kevin McBeath to languish in the less spectacular world of commerce where he can still be heard muttering quietly over money!

Jochen Schubert

Brilliant young German guitarist, JOCHEN SCHUBERT, came to Australia in 1970 to teach at the Melbourne Conservatorium of Music introducing classical guitar as a chief study subject.

He has played with many leading Australian musicians, such as Vernon Hill, Phillip Miechel, Leslie Miers, Henry Wenig and Leon La Gruta. He has made many solo appearances and premiered works of Wesley Smith and Brideoak at the opening season of the Svdney opera house.

His recordings include Felix Werder's *Banker In Melbourne* and Rodrigo's *Concerto De Aranjuez* with the Alan Lee Quartet.

Jochen Schubert lives in Kew with his charming wife and small daughter. The Schubert family has a menagerie of animals and their home is known as the "Kew Zoo".

Promotion – Nada Brozel

Figure 2.3 (Continued)

The cultural environment surrounding the ensemble also influenced their work. For example, Seal's *Attention Joe Brown! Who Stole the Melbourne Cup from the Grainger Museum?* (1972) and *Melbourne Media* (1973) reference the context of the Australian Percussion Ensemble's output in a city famed for its love of sporting events. Melbourne is acknowledged as the home of Australian Football League (known in the 1970s as Victorian Football League), and *Melbourne Media* is scored for six percussionists and a tape of a Victorian Football League radio broadcast. Although the practice of percussionists composing and performing their own works is now common, it was only beginning to emerge in global contemporary percussion circles in the 1970s as ensembles sought unique repertoire.

PRIME RIFF — Keith Humble

Prime Riff was commissioned by the A.P.E. for its current series of concerts. It is written for 6 percussionists and a pre-recorded tape. The material was prepared by a computer process and progresses from "non" pitched organisation to a pitched complex – it advances from "seeming" disorder to precision.

Ideally the work is spatial, however under tonight's performance limitations the instruments are organised to give an equivalent effect.

Prime Riff is the outcome of research processes evolved during the composition of both the Arcade series and Treatments 1-19.

STEREOPHONY — Paul Copeland

Stereophony is an attempt to demonstrate the existence of a line of demarcation which exists between electronic music and instrumental music.

The percussionist's part moves randomly from free imporvisation to strict note serialization. The rhythm is mainly aperiodic but, where the logic demands, a rhythmic cell is used to give coherence to the structure.

The electronic sounds are organically functional and exist for their own intrinsic value; nevertheless they are simultaneously dependent on, and independent from, the live sounds. This dual function creates for the listener a 'sound complex' which is continuously changing from 'concordant totality' to 'discordant non-associative sectionalisation'.

The electronic sounds were produced on an E.M.S. AKS synthesizer.

The piece was written for, and is dedicated to, John Seal.

IMAGES FOR CHRISTMAS — Helen Gifford
(to a poem by Peter Murphy)

for voice, electric guitar, small organ, celesta and percussion

Soloist: Kevin McBeath

The musical base and inspiration for this work is the old German chorale melody "Der du bist drei in Einigkeit" (Thou Who art Three in Unity) which probably had its origin in plainsong.

In this work, the melody appears as an inner voice canto fermo in three separate and self-contained segments on the organ, and fully stated or fragmented and sounding in different juxtaposition against each other in the tuned percussion.

Helen Gifford chose this very old sacred melody for her Christmas piece in preference to an actual Christmas carol. Most true carols in fact date from a much later period, there being few that survive from any time earlier than the fifteenth century. For Helen Gifford, this melody more truly represents the simple essence and piety of early religious belief that – with our current Christmas festivities – has now become submerged in the clamour and glitter of our seasonal celebrations.

PETER MURPHY is a young Melbourne writer who has recently completed the libretto for Helen Gifford's opera *Jo Being*. A collection of his poetry, *Escape Victim and Other Poems,* is due to be published shortly.

Peter Murphy says about *Images For Christmas* that it is "a poem that tries to come to terms with the sources of Christmas". The mood of the poem is at once sympathetic and critical and aims at a kind of emotive re-evaluation of the Christmas phenomenon.

INTERVAL

DIALOGUE — Siegfried Fink

Soloist: Jochen Schubert

Along with Christof Caskel, Siegfried Fink is one of Germany's exciting percussionists. He is a composer of some note, especially of works featuring solo percussion and as is the case tonight, percussion and guitar. Siegfried Fink and guitarist Siegfried Behrend to whom tonight's work is dedicated have an enviable duo reputation for the performance of new music featuring this combination.

TRIPTICON — Simeon Ten Holt

Simeon Ten Holt, one of the Netherlands' most advanced composers, has dedicated this work to HC Ten Berge.

The work is in three movements and explores the potentials of different tempos played either simultaneously or consecutively. The first and third movements are closely related and consequently the work could be considered to be in quasi-sonata form. The second movement is the most virtuosic of the piece and is comprised of episodic fragments of both varying tempos and silences.

Figure 2.3 (Continued)

The performance of works by percussionist-composers from within the ensemble was another aspect of the APE's programming that could be considered pioneering for its time, as the performance of works by established peers and colleagues was adventurous enough for many Australian chamber groups in the 1960s and 1970s. In the 1970s, international contemporary percussion groups such as Canadian percussion ensemble NEXUS (established in 1971) were beginning to compose and perform their own repertoire alongside improvisations, commissions and existing repertoire. As NEXUS had not made any recordings or toured Australia prior to APE's formation, it can be concluded that NEXUS and the APE reached this programming model independently. The 1970s are acknowledged as a watershed

Figure 2.4 John Seal with his wooden boxes for *Structures*, August 1974.

Source: Supplied by Wendy Couch. Used with permission.

period in global contemporary percussion with a major increase in output from artists around the world (Schick, 2006a: 67). That the APE, and every professional Australian percussion ensemble established during the same period, favoured this programming model suggests that Australian contemporary percussion music was operating with ideologies and creative processes similar to those of international percussionists of the same period.

Table 2.1 lists all of the Australian repertoire performed by the APE and highlights the variety of their repertoire during the five years the group was active. Contemporary sub-genres programmed by the group included electronic, experimental, improvisational, collaborative, graphic score and jazz-influenced works. Twenty-six of the 29 works shown in Table 2.1 were commissions and therefore represent a significant contribution to the emergent Australian percussion

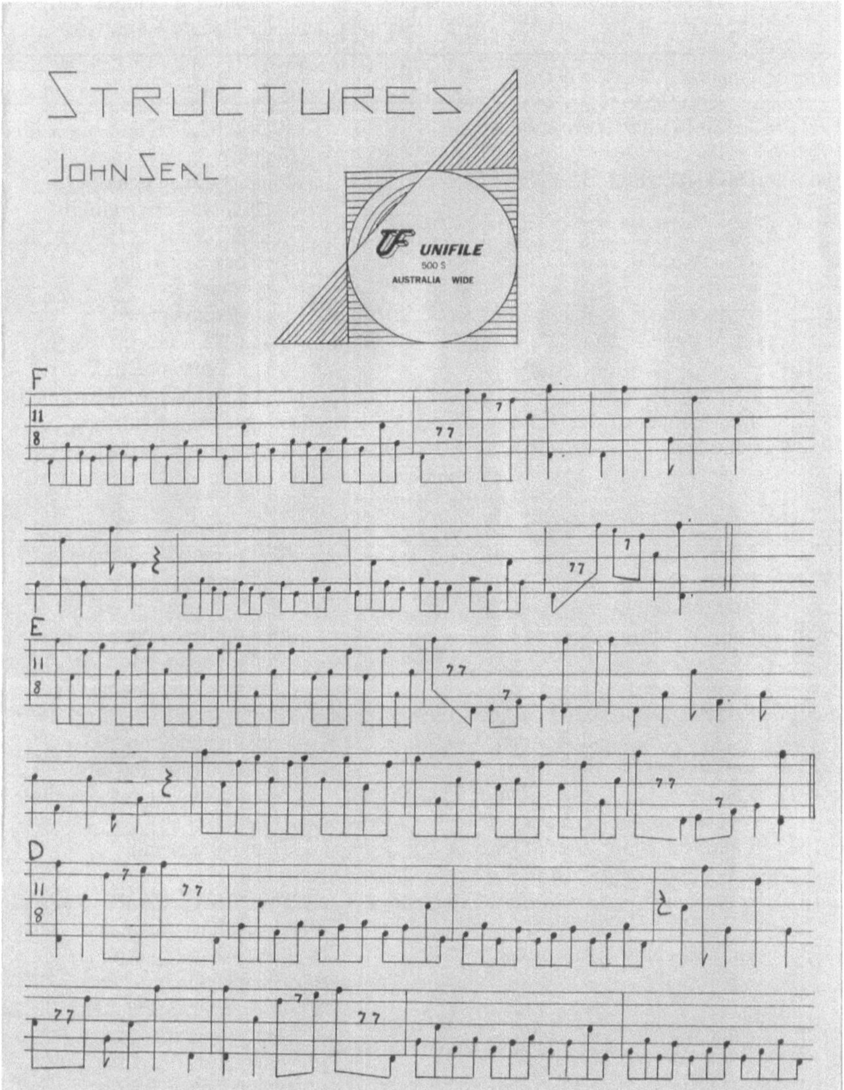

Example 2.1 Page one of *Structures* by John Seal.

Source: Score supplied by John Seal.

repertoire of the time. The Australian Percussion Ensemble recorded a small number of these works, which have contributed to the preservation of this repertoire. The most notable of these is a recording of repertoire by Ian Bonighton titled *Sequenza: And Other Works* (Bonighton, 1977), which included *In Nomine,*

Table 2.1 Australian repertoire performed by the Australian Percussion Ensemble. (Where the complete scores have been lost, the approximate dates of composition have been drawn from concert programs within Couch and Babaré's archives.)

Year	*Composer*	*Work*	*Instrumentation*
1911–1913	Percy Grainger	*Random Rounds**	Played by APE on piano with guests
1967	David Ahern	*Music for Nine: Nonet**	Flute, clarinet, piano, two percussion, two violins, viola, cello
1971	Felix Werder	*Oscussion**	Two percussion and synthesizers
1972	Felix Werder	*Percussion Play*	Solo percussion Mvt 1 Untuned percussion Mvt 2 Tuned percussion Mvt 3 Timpani (unperformed)
1972	Ian Bonighton	*Derivations III*	Six percussion and tape
1972	John Seal	*Attention Joe Brown! Who Stole the Melbourne Cup from the Grainger Museum?*	Tape, no live performers
1972	John Grant	*Exposition*	Four percussion and piano
1972	Peter Brideoake	*Flute*	Flute and two percussion
1972	Felix Werder	*Scherzi*	Jazz-influenced quartet of vibraphone, drum set, trumpet and double bass
1972	Ron Nagorcka	*Six Numbers Converging on an Animal*	Graphic score with free instrumentation
1972	Ian Bonighton	*Spectra*	Six percussion and tape
c. 1972	Jennifer Mackenzie	*Then What Is That Distant Land*	Poetry and chamber ensemble
c. 1972	John Seal	*That From Your Life Seems to Draw All Its Light*	Chamber ensemble
c. 1972	John Seal	*Yarra Bank Concerto*[2]	Text A'Beckett, music by Seal Musical theatre/chamber music
1973	Ian Bonighton	*In Nomine*	Percussion, organ and tape. Appears on *Sequenza: And Other Works*
1973	Felix Werder	*Banker*	One percussion, two synthesizers, guitar, piano. Appears on *Banker*
1973	Ron Nagorcka	*Caucasian Chalk Circle*[3]	Percussion, actors, singers
1973	Helen Gifford	*Images for Christmas*	Speaker, electric guitar, small organ, celesta and five percussion
1973	Robert Irving	*Sums*	Four percussion or of any instrument

Year	*Composer*	*Work*	*Instrumentation*
c. 1973	John Seal	*Melbourne Media*	Six percussion and tape of Australian Rules Football broadcast
c. 1973	David Bedford	*Variations on a Rhythm by Mike Oldfield*	Six percussionists playing 36 bottles, 18 flowerpots, 9 cymbals, 3 snare drums, 3 timpani, 12 squeaky toys, piano, tubular bells and vacuum cleaner
c. 1973	Felix Werder	*Toccata Raga*	Unknown
1974	Ian Bonighton	*1, 2, 3*	Three percussion
1974	Keith Humble	*Prime Riff*	Six percussion and tape
c. 1974	Paul Copeland	*Stereophony*	Percussion and electronics
1974	John Seal	*Structures*	Two or four percussion for any instrumentation
1977	Warren Burt	*Salon Music*	Off-stage piano and two xylophones
n.d.	Graham Hair	*Creation*	Canta for soprano, piano and five percussion
n.d.	Felix Werder	*Opus 131*	Two percussion, double bass and trumpet

* Existing work

Derivations III and *1, 2, 3*. Bonighton passed away in 1975 and this recording was made posthumously by Keith Humble, Douglas Lawrence, Rosslyn Farren-Price, Alex Grieve, the Australian Percussion Ensemble and the Melbourne String Quartet. As few Australian percussion recordings exist from the 1970s and few works by Bonighton have been recorded, *Sequenza* has become an important artefact of Australian compositional history.

International percussion practices in the 1970s frequently included explorations integrating acoustic and electronic textures and sounds. This trend was a major part of the practice of a number of expatriate composers in Melbourne who wrote for the APE, such as Warren Burt and Felix Werder. Both Burt and Werder are key figures in the history of experimental and contemporary music composition and electroacoustic performance practice in Australia. Burt cites Werder's *Oscussion* (1971) for two percussion and synthesizers as an early Australian example of experimentation between acoustic percussion and electronic instruments, and this work has the additional distinction of being one of the earliest Australian works for percussion (Burt, 2017). It is pertinent to note that *Oscussion* was composed in the same year as the sextets composed by Peter Sculthorpe and Barry Conyngham for Les Percussions de Strasbourg. Later premièred by the APE, *Oscussion* was composed prior to the foundation of an ensemble that might perform it, mirroring the circumstances of early percussion works in North America and Europe, as discussed in Chapter One.

As Table 2.1 demonstrates, the Australian Percussion Ensemble was prolific in the development and performance of new repertoire during the years it was operational, and in their methods of presenting concerts. Their work was recognised nationally in the form of invitations to perform at the Festival of Perth (now the Perth Festival) and David Ahern's Sydney-based contemporary music project, AZ Music Series. However, the short-term impact of the APE was limited as the group dispersed after approximately five years. The logistics of running an independent percussion ensemble in combination with a lack of support from the wider concert-going public were primary contributors to the group's short existence. Although numerous new works were created, further development of many of these works was limited by the fact that the majority were only presented once in performance, which in turn prevented them from staying in the repertoire. There were various reasons for presenting such works only once, including audience saturation and logistics. Many of the works written for the APE were scored for enormous percussion set-ups, and although these set-ups occasionally shared instruments between players, negotiating such large set-ups presented some difficulties. An example can be found in *Spectra*, a 1972 commission from James Penberthy in four movements titled *Very Rumbustious, Very Delicate, Very Gymnastic* and *Very Awesome. Spectra* called for timpani, tambourine, glockenspiel, xylophone, vibraphone, jazz drum kit, bongos, bells, rhythm sticks, bass drum, triangle, temple blocks, cowbell, bird warbler, siren, guiro, large gong, small gong, cymbals, glasses, Indian bells, small Japanese cymbals, snare drum, beer bottles and tape.

The APE's activities marked the beginning of a professional contemporary percussion scene in Australia, and just as François' short tenure in Melbourne had been a loss, the cessation of the APE's activities was an additional blow to the Melbourne artistic landscape. However, the imaginations of Australian musicians with regard to the creation of percussion music had been ignited, and composers continued to develop percussion repertoire for Seal and others beyond the formal ending of the APE. An example can be found in Dieter Boytek's *Three Sound-Molecules* (1979) for two percussionists. Although the work of the Australian Percussion Ensemble has now largely faded from memory, it cannot be disputed that their work made considerable contributions to the emergent Australian percussion repertoire and identity of the time. Gradually, the repertoire generated by the APE is beginning to be recognised by a new generation of Australian percussionists seeing links to their local history. Examples of this recognition include the re-working of APE commission *Salon Music* (1977) by Warren Burt for Speak Percussion's performance in 2009, and the reconstruction and performance of two works from the Three Creative Recitals series by Louise Devenish in 2015: Keith Humble's *Prime Riff* (1974) and John Seal's *Structures* (1974).

The Victorian College of the Arts

Beyond the activities of the APE, growth in the wider Australian musical landscape in the 1970s had positive implications on the nascent Australian percussion scene, particularly in tertiary contemporary music education in Melbourne. In the early 1970s, percussion was initially taught privately or as a single study at Melbourne

University Conservatorium of Music as previously mentioned. The line of Australian percussion teachers in Melbourne prior to the 1970s can be traced from John Seal back to his private teachers and later colleagues Jean-Charles François and Bill Fleming. Fleming was a freelance professional percussionist and drummer who performed with the MSO, the ABC Showband, in theatres and with dance bands. According to Seal, Fleming studied drums and some percussion with Jack Soloman, an English musician who migrated to Australia in the early 1900s (Seal, 2012). Following François' departure from Melbourne, Seal began teaching students at the Melbourne Conservatorium, including future APE members Couch and Babaré. By the end of the 1970s, two additional music institutions were established in Melbourne that would strongly influence the direction of percussion education in twentieth-century Australia: La Trobe University and the Victorian College of the Arts.[4] La Trobe University was established in 1975, and the music department later specialised in experimental music, electronics and improvisation. When John Hopkins founded the Victorian College of the Arts (VCA) School of Music in 1974, structured tertiary opportunities arose for young percussionists for the first time in Melbourne. Hopkins engaged Barry Quinn (b. 1936) to assist in the establishment of a percussion course, and, significantly, three percussionists were part of the first intake of students. This was the first time that percussionists were included in the first student cohort at a tertiary institution in Australia. Quinn, shown in Figure 2.5, was the first full-time percussion teacher in Melbourne and was central to percussion education in this city between 1974 and 2000.

As formal training in Australia was not available during his formative years, Quinn was part of the generation of percussionists (along with John Seal and Richard Smith) who became professional through a combination 'learning on the job' and ad hoc private lessons. Quinn began with drum set lessons from his uncle, renowned jazz drummer Billy Hyde, followed by percussion lessons from MSO principal percussionist Ernest 'Ernie' Lighton at the age of 17. New Zealander Lighton had come from the tradition of army bands and taught rudimentary percussion technique founded in this genre. Quinn began a career as a dance band and radio drummer in 1958 at the age of 22. When the Billy Hyde Drum Clinic opened in 1962, Quinn joined Garry Hyde as a tutor until 1965, when he was offered the position of drummer aboard *The Patrice*, a cruise ship bound for London in April of the same year. Shortly after arriving in London, Quinn approached BBC Symphony Orchestra percussionist Gilbert Webster, who also taught percussion at the Guildhall School of Music, for private percussion lessons. A variety of freelance work followed for Quinn, including work with the BBC Symphony Orchestra, pit orchestras for musicals and contemporary chamber ensemble The Fires of London, bringing Quinn into contact with some of the leading figures in European contemporary music of the time over the next seven years.[5] Quinn returned to Australia when a position became available with the MSO in 1971, and speaks of the audition process in the early 1970s:

> I kept in touch with Ernest Lighton. We used to write letters, he was a very nice man, and he mentioned that he was retiring and that the orchestra would be looking for a principal percussionist. I wrote and expressed interest to the

Figure 2.5 Publicity photograph of Barry Quinn, c. 1975.

Source: Image supplied by Barry Quinn.

> ABC and they said, 'yes, your credentials look good'. So I sent a tape and they said, 'Look, you would get this job, but we can't give it to you, you're not here'. At that time, I had got married, and my wife was pregnant with our first child, so we had to make up our minds quickly. So, we got on board a ship and came over. The day that the ship arrived, I got off and phoned from the wharf to the ABC, and said, 'I'm here'. They said, 'Well, we're just having a morning break, but we're going on Christmas holidays after lunch. So if you want to do an audition, you should come to the studio straight away'. So I went over to the studio. I hadn't played in six weeks! That's the only audition I've ever done for a percussion job.
>
> (Quinn, 2013)

In January 1972, Quinn took up the position of principal percussionist with the Melbourne Symphony Orchestra, a position he held for seven years. In addition to his work with the MSO, Quinn established percussion rental company Percussion Services in 1973 (renamed Myriad Percussion in 1980) and performed with chamber music ensembles such as the Australian Contemporary Music Ensemble (ACME). Founded by Keith Humble and Don Banks, ACME's programming featured Australian music by Humble, Banks and their colleagues such as Larry Sitsky and their repertoire includes some of the early Australian works including contemporary percussion in a chamber music ensemble setting.

In 1975, Quinn began teaching percussion at the VCA. As part of a Bachelor of Music degree, percussion tuition at the VCA began with a focus around a single lesson per week. Initially, the VCA percussion program was designed to prepare students for orchestral careers, and this was the primary focus of many performance students. In addition to the tertiary course, the VCA secondary school included percussion students and the direct transition from the school aided rapid growth in student numbers. Quinn was the primary teacher for three years, and was joined in 1978 by newly appointed MSO timpanist Larry Wilson. Wilson had studied with respected timpanist Saul Goodman in New York and migrated from America to take up the MSO position. By this time, there were sufficient numbers to form a student percussion ensemble, and the ensemble served as an introduction to contemporary percussion repertoire for many students (Pye, 2012). The exploration of contemporary ensemble repertoire became a secondary focus of percussion study at the VCA (Pye, 2012), further enhanced by activities surrounding the music course. VCA Director John Hopkins placed great value on interaction between students, staff and external professional artists, and from the late 1970s, students and faculty members also performed in a number of contemporary and world music projects at the VCA and in other parts of Melbourne, including Richard David Hames' contemporary ensemble the Victorian Time Machine (see Fox, 2002: 159–163) and the Monash University Gamelan Ensemble. Quinn cites Hopkins as the driving force behind the interaction between visiting contemporary percussionists, students and faculty at the VCA in the 1970s and 1980s, observing 'John Hopkins was in touch with what was going on. He would come and say: [French percussionist] Sylvio Gualda is coming to Melbourne. I think we should have him visit [the VCA]. Can you arrange it?' (Quinn, 2013). David Pye and Robert Clarke constituted the second intake of percussionists in 1977, Pye states that the experiences with visiting guest artists at the VCA were vital to his contemporary percussion development because they exposed him to the work of active Australian and international composers and instrumentalists, graphic scores and improvisation. In the 1970s, two North American visitors had a significant impact on VCA percussionists – composer Bill Fontana and timpanist George Gaber – before Gualda's visit in the 1980s.

An experimental sound artist who moved to Australia in 1974, Bill Fontana (b. 1947) worked for the Australian Broadcasting Corporation 'recording what Australia sounded like' (Fontana, 2017) between 1974 and 1978. In 1976, Fontana began creating large, site-specific installations also known as sound sculptures. Inspired by the site of the installation, these sound sculptures include museums, galleries, public spaces and public infrastructure such as bridges in Sydney, Paris, London, San Francisco, Tokyo, Berlin and New York. In addition to composing and creating sound sculptures in Australia, Fontana ran a contemporary concert series in Melbourne (1976–1978) that hosted a plethora of contemporary composers, among them John Cage (Burt, 2007). He also gave workshops and instigated experimental projects at RMIT University (formally known as Royal Melbourne Institute of Technology), La Trobe University and the VCA. Pye cites Fontana as 'pivotal in my development, particularly in terms of listening to environmental sound' (Pye, 2012). This is reflected in the high number of environmentally influenced site-specific works in Pye's catalogue as a composer and director. These include *Their Kind of Moon* (Pye, 2007) for marimba, woodblocks and tape of pre-recorded frog calls from the South-West of Australia, and *Karakamia* (2004), a 45-minute chamber ensemble work designed to draw attention to and celebrate the work being done at the Karakamia Wildlife Sanctuary.

George Gaber (1916–2007) was an orchestral timpanist, studio and freelance musician who developed a strong percussion curriculum and program at Indiana University between 1960 and 1986 (Kite, 2017). His reputation for excellence as a percussion educator led to an invitation from John Hopkins to present clinics at universities in Perth, Melbourne and Adelaide in 1975. Given that there were few opportunities to learn percussion at the time of Gaber's visit, it is perhaps unsurprising that the two-week residence of an international percussion specialist in Melbourne drew percussionists from cities not on Gaber's itinerary. These included Sydney Symphony Orchestra (SSO) percussionist Colin Piper, who remembers the inspirational effect Gaber had on professionals and students alike during his visit (Piper, 2013). The positive outcomes of Gaber's visit highlighted the importance of connecting with established international percussionists on the rare occasions they visited Australia. As Australian percussion activity continued to grow, it became common for percussionists to travel to neighbouring cities to be involved in intensive projects with international percussionists in residency at institutions, and this pattern continued until the 2000s. An example of another intensive project can be found in French percussionist Sylvio Gualda's (b. 1939) visits to Melbourne in 1981 and 1983.

Gualda's performances in Melbourne set a benchmark for solo percussion performance in the 1980s in the same way that Les Percussions de Strasbourg set a benchmark for percussion ensemble performance in the 1970s, and served to continue the influence of French percussion practices on Australian activity. Like Les Percussions de Strasbourg, Gualda brought some of his own percussion instruments such as contrabass toms custom-made for Iannis Xenakis's *Psappha* (1975), the first of two percussion soli composed for Gualda by Xenakis that have become staples of the international percussion repertoire. Additionally, Gualda's Melbourne performances of cutting-edge percussion solo repertoire included a number of pieces from his 1978 recording *Percussion Vol 2* (Gualda, Xenakis, Carter, & Nguyen, 1978) such as Elliott Carter's *Eight Pieces for Four Timpani*. With much of the works performed less than ten years old, Gualda's programs consisted of groundbreaking international works. This tour introduced this repertoire to Australian audiences while it was still considered new (Neville, 2013). The schedule of his second residency at the VCA is shown in Figure 2.6.

In 1982, Quinn's role as Head of Percussion was formalised at the VCA, and the percussion department activity increased. In 1984, Quinn spent ten weeks travelling to various international conservatoria, publishing houses and the studios of professionals in North America and Europe in an effort to increase his

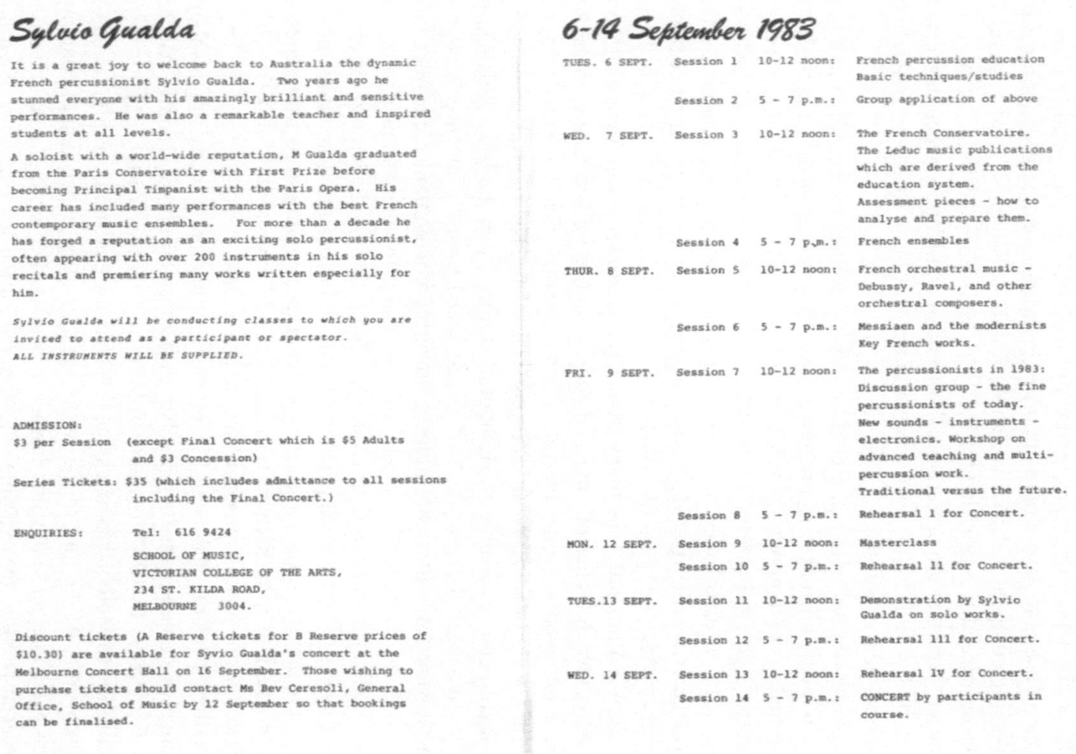

Sylvio Gualda

It is a great joy to welcome back to Australia the dynamic French percussionist Sylvio Gualda. Two years ago he stunned everyone with his amazingly brilliant and sensitive performances. He was also a remarkable teacher and inspired students at all levels.

A soloist with a world-wide reputation, M Gualda graduated from the Paris Conservatoire with First Prize before becoming Principal Timpanist with the Paris Opera. His career has included many performances with the best French contemporary music ensembles. For more than a decade he has forged a reputation as an exciting solo percussionist, often appearing with over 200 instruments in his solo recitals and premiering many works written especially for him.

Sylvio Gualda will be conducting classes to which you are invited to attend as a participant or spectator.
ALL INSTRUMENTS WILL BE SUPPLIED.

ADMISSION:
$3 per Session (except Final Concert which is $5 Adults and $3 Concession)

Series Tickets: $35 (which includes admittance to all sessions including the Final Concert.)

ENQUIRIES: Tel: 616 9424
SCHOOL OF MUSIC,
VICTORIAN COLLEGE OF THE ARTS,
234 ST. KILDA ROAD,
MELBOURNE 3004.

Discount tickets (A Reserve tickets for B Reserve prices of $10.30) are available for Syvio Gualda's concert at the Melbourne Concert Hall on 16 September. Those wishing to purchase tickets should contact Ms Bev Ceresoli, General Office, School of Music by 12 September so that bookings can be finalised.

6-14 September 1983

TUES. 6 SEPT.	Session 1	10-12 noon:	French percussion education Basic techniques/studies
	Session 2	5 - 7 p.m.:	Group application of above
WED. 7 SEPT.	Session 3	10-12 noon:	The French Conservatoire. The Leduc music publications which are derived from the education system. Assessment pieces - how to analyse and prepare them.
	Session 4	5 - 7 p.m.:	French ensembles
THUR. 8 SEPT.	Session 5	10-12 noon:	French orchestral music - Debussy, Ravel, and other orchestral composers.
	Session 6	5 - 7 p.m.:	Messiaen and the modernists Key French works.
FRI. 9 SEPT.	Session 7	10-12 noon:	The percussionists in 1983: Discussion group - the fine percussionists of today. New sounds - instruments - electronics. Workshop on advanced teaching and multi-percussion work. Traditional versus the future.
	Session 8	5 - 7 p.m.:	Rehearsal 1 for Concert.
MON. 12 SEPT.	Session 9	10-12 noon:	Masterclass
	Session 10	5 - 7 p.m.:	Rehearsal 11 for Concert.
TUES.13 SEPT.	Session 11	10-12 noon:	Demonstration by Sylvio Gualda on solo works.
	Session 12	5 - 7 p.m.:	Rehearsal 111 for Concert.
WED. 14 SEPT.	Session 13	10-12 noon:	Rehearsal 1V for Concert.
	Session 14	5 - 7 p.m.:	CONCERT by participants in course.

Figure 2.6 Schedule of events for Sylvio Gualda's 1983 visit to Melbourne.

knowledge of international teaching practices. Interestingly, it was during this tour that he began to realise that although Australian percussionists were more aware of international practices than they had been 20 years earlier, the reverse was not yet true. In a notebook from that trip Quinn wrote 'the only [Australian] piece anybody has heard of over here is David Lumsdaine's *Kangaroo Hunt'* (Quinn, 2013). Upon his return, percussion at the VCA continued to grow. The percussion ensemble gave rise to other groups whose members were drawn from its ranks, including educational ensemble Rainbow Rhythmix, and Festival Percussion, an ensemble assembled for the 1991 Melbourne Festival. As more VCA graduates moved into the profession, the seeds of contemporary percussion activity sown by the APE in the 1970s began to be built upon by a new generation of professional percussionists.

The percussionists who completed their VCA studies in the 1970s and began careers in music all started as orchestral percussionists, although a number have since turned to contemporary percussion or other fields. In the 1980s, in addition to pursuing orchestral careers, a number of percussionists entered the profession as freelancers or established their own ensembles. Among the graduates of this time was Peter Neville (b. 1963), who became a leading figure in contemporary percussion in Melbourne in the 1980s and succeeded Quinn as Head of Percussion at the VCA in 2000. Like many freelance percussionists in Australia, Neville began his career in orchestras, musical and visual theatre, dance and opera, however he maintained a strong interest in contemporary chamber music. Together with guitarist Daryl Buckley and composer Liza Lim, Neville was involved with the New Audience Ensemble at Melbourne University in the late 1980s. Despite jokingly referring to the group as the No Audience Ensemble, the three recognised that there were few other ensembles performing contemporary chamber music repertoire at the time, and they established Elision to address this gap in 1986. Elision enabled the trio and their colleagues to perform and compose experimental new music that challenged performers and audiences alike, and now includes 16 ensemble members. Elision's repertoire includes commissions from renowned composers such as Lim, Richard Barrett, Michael Finnessy and Chris Dench, and interpretation of Franco Donatoni's music has become a hallmark of the ensemble. Neville's performances with Elision were influential in increasing awareness of a range of contemporary approaches to percussion composition in Australia during the 1980s. In 1997, the ensemble moved its base to Queensland before later returning to Melbourne, and since this time the ensemble has premièred over 200 new works. At the turn of the twenty-first century, experimental contemporary percussion began to develop further in Melbourne, through solo and ensemble projects. Neville, for example, was the inspiration for a number of virtuosic solo percussion works that broached new sonic territory. Many have since entered the Australian percussion canon, such as Andriàn Pertout's *Exposiciones* (2007) for glockenspiel and tape.

Since 2000, the trajectory of exponential growth in Australian contemporary percussion practices that commenced in the 1970s has continued. The influence of other performing arts and conceptual art has appeared in unique percussion projects that incorporate unusual or specialised instruments requiring development of new performance practices. Some of this activity, such as the Playing Fields quartet and Virtuosic Visions recital series feature individuals embedded in centres of activity outside Melbourne, and thus are discussed in Chapter Six rather than in the pages that follow. Others have remained in Melbourne. Speak Percussion, for example, has emerged as an internationally recognised leader in the fields of experimental, interdisciplinary and contemporary musics, and has been influential in navigating new directions in twenty-first-century Australian percussion music. Although their most significant output falls beyond the time period examined in this monograph, it is, however, representative of new directions in Australian contemporary music and thus serves to introduce concepts that feature in later chapters of this book.

Speak Percussion

Founded in 2000 by Minako Okamoto, Justin Marshall, Rory McDougall, Phil Binotto and Eugene Ughetti, Speak Percussion grew out of the 1990s student cohort at the Victorian College of the Arts. Notably, Speak Percussion was one of the first professional Australian ensembles that did not emerge from the percussion section of a state symphony orchestra, evidencing growth in the size and scope of percussion activity in Australia. Further, this has enabled the ensemble to follow a different path from their contemporaries and predecessors as the ensemble's members were not also employed by a major performing arts organisation. Since 2003, Eugene Ughetti has been the ensemble's Artistic Director, and has focussed the ensemble on the commissioning and performance of over 100 new works. Cultivating long-term relationships with composers in the development of this repertoire has been crucial to the ensemble's engagement with the surrounding musical landscape. The focus on composer-performer collaboration with Australian composers is discussed by Ughetti thus:

> We're acutely aware of where we came from and where we continue to live. Which means that we are Australian and we are also Melburnians. What's important about that is that we have an obligation to our community to develop and to enrich our local culture. Local culture can literally be the very close community, all the way out to the whole country. So what we've done is prioritised that kind of work. The vast majority of our work is Australian repertoire. That doesn't come out of nationalistic pride, it comes out of this genuine commitment and sense of responsibility to propagate Australian culture and to enrich our immediate community.
>
> (Ughetti, 2014)

Significant amongst these long-term collaborations with Australian composers are works by Liza Lim, Anthony Pateras and Thomas Meadowcroft, that are now significant in the international canon. For example, between 2004 and 2011, Speak Percussion commissioned, performed and recorded a range of works from

Anthony Pateras, from solo works to large ensembles such as sextet *Refractions* (2008) and twelvetet *Flesh and Ghost* (2011). Speak Percussion has also championed Thomas Meadowcroft's music. Although Meadowcroft has been an expatriate for most of his career, Ughetti describes Meadowcroft's works as being 'the most Australian-sounding percussion music I have ever heard' (Ughetti, 2017), largely due to their instrumentation that includes a range of everyday urban objects and sounds. These works often feature objects common in domestic music making such as plastic classroom recorders and tapes, heard in *The Great Knot* (2011).

In recent years, Speak Percussion projects have pushed at the edges of contemporary percussion practice, engaging in interdisciplinary collaborations linking percussion with architecture, kinetics or gastronomy, and integrating major influences from digital technology such as lighting systems and video. The design and development of new instruments and performance practices has become a key feature of the ensemble's output. An early introduction to interdisciplinary collaboration and instrument making came via Elaine Miles' *The Glass Percussion Project* (2006–2010), which required Ughetti and Speak Percussion Artistic Associate Matthias Schack-Arnott to sound an installation of over 2500 glass objects in performance. Also evident in Speak Percussion's output is the twenty-first-century trend in work identifying as percussion music but without using recognisable percussion instruments or techniques. This has been termed post-percussive practice by Håkon Stene (Stene, 2015: 2), and will be discussed further in Chapter Seven. An example of Speak Percussion's work in this area can be found in *TRANSDUCER* (2013), wherein the central instruments are microphone and loudspeakers (see detailed analysis in Ughetti, 2017). A single tam-tam is the only recognisable percussion instrument used in the work, however it is not used conventionally. The tam-tam is a visible presence on stage for the entirety of the 50-minute work, but when it is finally approached it is not struck or otherwise physically sounded by the performer. Instead, it is sounded via feedback between a microphone and speaker that activates the instrument sympathetically. *TRANSDUCER* is the first of a number of major post-percussive projects by Speak Percussion, and was followed by *Fluorophone* (2015), *Annica* (2016) and *Assembly Operation* (2017).

This chapter shows that the foundations of a thriving contemporary percussion scene were laid in Melbourne by a diverse range of performers, composers and pedagogues, and that this work was influenced by international trends in experimental music and French contemporary percussion practices. Collaboration between composers and performers steered the development of an emergent body of Australian repertoire for the Australian Percussion Ensemble in the 1970s, who led professional contemporary percussion activity in Melbourne between 1972 and 1977. Activity within and adjacent to tertiary institutions became central in the 1980s and 1990s as new opportunities for percussion study and performance arose. Towards the turn of the century, new voices in experimental percussion practices began to emerge in Melbourne through the activities of Peter Neville and later through Speak Percussion. Although the activity in Melbourne was the first professional activity in Australia in the early 1970s, it was also during this decade that percussion practices began to emerge in Sydney through the activities of Synergy Percussion. The circumstances that saw Sydney move forward as a hub for contemporary percussion in the late 1970s and 1980s are examined in Chapter Three.

Notes

1 *Prime Riff* is listed in John Whiteoak's catalogue of Humble's archives held in the National Library of Australia in Canberra with the commission date of circa 1976. The APE program shown in Figure 2.2 serves to update this estimated composition date.

2 This work was performed in the ensemble's inaugural concert in 1972. The program also contained what appear to be text-based works or poems titled *Scenes of the City* and *Poems of the City*.

3 Not to be confused with Helen Gifford's composition of incidental music for the Melbourne Theatre Company's production of Brecht's *The Caucasian Chalk Circle* (c. 1970). In personal communication between this author and Nagorcka, the composer stated that this work was performed by the APE, but was written 'with no particular percussionists in mind'. Nagorcka composed a small number of works utilising percussion instruments in this way. An example is *Atom-Bomb Trilogy* (1977) for percussion, toy instruments, tape recorders and clavichord. A detailed analysis of this work can be found in *Experimental Music in Melbourne: 1975–1979* (Fox, 2002: 75–88).

4 Monash University's existing program also included music, however this was primarily focussed on ethnomusicology at this time.

5 Founded as The Pierrot Players, The Fires of London was directed by contemporary composers Peter Maxwell Davies and Harrison Birtwistle between 1965 and 1987. This influential ensemble premièred numerous contemporary chamber music works and toured extensively throughout Europe. The Pierrot Players' first program featured Arnold Schoenberg's *Pierrot Lunaire* (1912) for flute, clarinet, violin, cello and piano. The Pierrot Players subsequently adapted the instrumentation of that work by adding percussion to create their core ensemble. This instrumentation is known as a 'Pierrot ensemble' or 'Pierrot plus' and has become a common foundation for numerous contemporary chamber ensembles around the world, including Eighth Blackbird (USA), Decibel New Music (Australia) and Uusinta Lunaire (Finland).

References

Appleby, R. 2012. *Women of Note: The Rise of Australian Women Composers*. Fremantle: Fremantle Press.

Bonighton, I. 1977. *Sequenza: And Other Works*. Move Records MS3016–2LP.

Burt, W. 2007. "Some Musical and Sociological Aspects of Australian Experimental Music." Resonate Magazine, 31 July. www.australianmusiccentre.com.au/article/some-musical-and-sociological-aspects-of-australian-experimental-music

Burt, W. 2017. "Felix Werder and the Accidental Australian Adventure." Accessed December 7 2017. www.dramonline.org/albums/felix-werder-the-tempest/notes

Fox, R. 2002. "Experimental Music in Melbourne: 1975–1979." Master of Music diss., Monash University.

Fontana, B. 2017. "Resoundings." Accessed December 7 2017. http://www.resoundings.org/Pages/Resoundings.html

Gualda, S., I. Xenakis, E. Carter, and T. D. Nguyen. 1978. *Percussion Vol 2*. LP. Erato.

Kite, R. 2017. "PAS Hall of Fame: George Gaber." Accessed December 7 2017. www.pas.org/about/hall-of-fame/george-gaber

Piper, C. 2013. "The Early Days of Synergy." *DRUMscene*, 71: 102–103.

Schick, S. 2006a. *The Percussionist's Art: Same Bed, Different Dreams*, 1st ed. New York: University of Rochester Press.

Stene, H. 2015. "This Is not a Drum." DMA thesis. Norwegian Academy of Music. http://hdl.handle.net/11250/2379520

Sykes, J. 1973. "Scrounging Nice Sounds." *Sydney Morning Herald*, 24 August.

Ughetti, E. 2017a. "TRANSDUCER: Microphones and Loudspeakers as Musical Instruments." *Contemporary Music Review*, 36 (1–2): 48–63. doi:10.1080/07494467.2017.1368177

Selected scores cited

Burt, W. 1977. *Salon Music*. Melbourne: Self-published.

Carter, E. 1950/1966. *Eight Pieces for Four Timpani*. Minnesota: Hal Leonard.

Humble, K. 1974. *Prime Riff.* Melbourne: Self-published.

Nagorcka, R. 1972. *Six Numbers Converging on an Animal*. Sydney: Australian Music Centre.

Pye, D. 2004. *Karakamia*. Perth: Self-published.

Pye, D. 2007. *Their Kind of Moon*. Perth: Self-published.

Seal, J. 1974. *Structures*. Melbourne: Unpublished.

Stockhausen, K. 1959. *Nr. 9 Zyklus*. London: Universal.

Werder, F. 1971. *Oscussion*. Melbourne: Self-published.

Xenakis, I. 1975. *Psappha*. Paris: Editions Salabert.

Personal communication

Couch, W. 2014. Live Interview with Louise Devenish, 10 November. Melbourne, Victoria.

François, J-C. 2012. Email Interview with Louise Devenish. Paris, France.

Neville, P. 2013. Live Interview with Louise Devenish, 31 July. Melbourne, Victoria.

Piper, C. 2013. The Early Days of Synergy. *DRUMscene*, 71: 102–103.

Pye, D. 2012. Live Interview with Louise Devenish, 24 September. Fremantle, Western Australia.

Quinn, B. 2013. Live Interview with Louise Devenish, 23 July. Moorabbin, Victoria.

Seal, J. 2012. Skype Interview with Louise Devenish, 18 December. Melbourne, Victoria.

Seal, J. 2013. Email Communication with Louise Devenish, 20 March.

Ughetti, E. 2014. Skype Interview with Louise Devenish, 27 June. Melbourne, Victoria.

Ughetti, E. 2017b. Personal communication with Louise Devenish, 4 December. Melbourne, Victoria.

3 Contemporary percussion ensembles in Sydney

This chapter examines contemporary percussion in Sydney in the 1970s and 1980s, with a particular focus on the work of Michael Askill and the ensemble known today as Synergy Percussion. Askill (b. 1952) is arguably the most important figure in the genesis of Australian contemporary percussion music and his work as a performer, composer, director and collaborator continues to influence contemporary musicians today (Cleworth, 2013; Edwardes, 2013; Lagos, 2013; Piper, 2013; White, 2012). In the 1970s and 1980s, Sydney became a hub of contemporary percussion activity, largely due to the presence of Synergy Percussion. The emphasis placed by this ensemble on composer-performer collaboration produced a large body of repertoire that became the foundation of an Australian percussion canon. Key collaborations with composers and performer-composers are presented, including examination of their wider influence at a time when Australia's cultural identity was emerging on an international stage. Although tertiary training programs in Sydney developed later, the activities of this professional ensemble and their founding director Michael Askill resonated with percussionists and students around the country, and the ensemble's influence extends beyond ensemble repertoire to solo contributions and interdisciplinary collaboration. Examination of Synergy Percussion is followed by a discussion of contemporary Japanese-Australian drumming ensemble Taikoz. The chapter concludes with a brief examination of activities by Karak Percussion, Daryl Pratt and Claire Edwardes and a detailed exploration of the output of Graeme Leak.

Originally from Adelaide, Michael Askill was one of the many students who benefited from percussion lessons with Richard Smith. Askill was introduced to Smith by his father in 1965 at 14 years of age and consequently began taking private lessons with Smith in orchestral percussion, ragtime xylophone and rudimental drumming. Following, he began working in orchestras in Adelaide, Melbourne and Tasmania on a casual basis. Askill's training with Smith was directed towards an orchestral career and in 1971 he gained a position with the Australian Elizabethan Theatre Trust Orchestra (AETT) in Sydney. In the same year, Les Percussions de Strasbourg toured Australia and Askill attended these concerts. Although familiar with their repertoire from the Vingtième Siècle Series records he had purchased from the Australian World Record Club, Askill reflects that seeing them perform live had a strong impact and that he 'had never seen a group play their kind of repertoire so well' (Askill, 2013a). During their 1971 tour, Askill was introduced to Les Percussions de Strasbourg through colleagues in the AETT orchestra who had worked overseas. This introduction started a dialogue between Askill and the Strasbourg percussionists. Les Percussions de Strasbourg subsequently invited Askill to study at their home base at Le Conservatoire de Strasbourg. Although already working professionally, Askill had not undertaken formal tertiary training in Australia. Askill readily accepted the offer as the opportunity held a dual appeal: Strasbourg boasted a vibrant arts scene that could potentially enable him to continue to perform professionally while studying, and as a number of Les Percussions de Strasbourg musicians held orchestral percussion positions with the city's two orchestras, he would be able to combine both contemporary and orchestral performance learning and opportunities. Thus, Askill commenced a one-year diploma with Les Percussions de Strasbourg director and L'Orchestre Philharmonique de Strasbourg timpanist Jean Batigne in 1972.

During the course of his studies, Askill was able to engage with percussion activity at the forefront of the international contemporary music scene, work with composers such as Olivier Messiaen and Tona Scherchen, and become familiar with new international solo and ensemble repertoire of the time. Shortly after his return to Australia in 1973, Askill was invited to join Les Percussions de Strasbourg as a core member of the sextet. This prestigious invitation would have provided Askill with the unique opportunity to continue working at the cutting edge of international percussion; however, Askill declined the invitation in order to continue to live and work in Australia. Askill's decision later proved to be significant for Australian percussion, when he became director of The Sydney Percussion. Askill reflects that he 'returned from France with the intention of forming a percussion group and . . . the formation of Synergy Percussion was directly influenced by my connection to and involvement with Les Percussions de Strasbourg' (Askill, 2013b).

The Sydney Percussion (est. 1974)

The foundation of The Sydney Percussion in 1974 represents a pivotal event in Australian percussion history. Occasionally appearing under the name The Sydney Percussions, The Sydney Percussion would later become known as Synergy Percussion; a prolific driving force in Australian contemporary music for four decades. Like Les Percussions de Strasbourg, the foundation of the group was suggested by a composer colleague of the original percussionists. This composer was David Ahern, who also facilitated performance opportunities for the Australian Percussion Ensemble (Piper, 2013). Ahern had studied composition with Nigel Butterley and Richard Meale in Australia, before travelling to Europe in order to work with Karlheinz Stockhausen in Germany and Cornelius Cardew in England. When Ahern returned to Australia in 1970, he established AZ Music (Knowles, 2009: 13). AZ Music's activities included an experimental performance ensemble, an improvisation ensemble named Teletopa and a concert series that provided numerous opportunities for composers, experimental artists and performers, even if the work performed in this series wasn't always easily digested by those in attendance. James Murdoch writes of 'stormy periods of outrage and scandal' surrounding Ahern's AZ concert series, citing very public 'fisticuffs, slamming of piano lids by irate administrators and mass walk-outs' as regular occurrences (Murdoch, 1975: 7). Many AZ performances took place at the Cell Block Theatre in Darlinghurst, Sydney. The Cell Block Theatre was known for its willingness to house experimental performances from all art forms and the theatre itself played a role in Australia's contemporary music history of the time. This has been wonderfully documented by Deborah Beck, to whom composer Peter Sculthorpe reported: 'I can't think of a recital hall in Australia that was so important in the history of music making' (D. Beck, 2011: 106). Ahern was particularly supportive of percussionists and was involved with many of the earliest Australian percussion performances at this venue and elsewhere. Ahern encouraged Askill, then SSO percussionist Colin Piper and two largely self-taught freelance percussionists Ian Bloxsom and Ron Reeves, to form a percussion ensemble to perform as part of the first concert in his 1974 AZ Music Series (Piper, 2013). Piper acknowledged the significance of Ahern in the foundation and continuation of Synergy Percussion:

> He had certainly planted the seed of a permanent group by giving us the opportunity of playing together for the first time as an ensemble. This was a new experience for Ian, Ron and myself, though of course Michael had already had the experience of working with Les Percussions de Strasbourg. It was also around this time that David gave us a score of Steve Reich's *Drumming*, suggesting that it too might be a good piece to perform.
>
> (Piper, 2013)

Percussionist Colin Piper commenced his tertiary music training as a pianist at the Sydney Conservatorium in 1965. At this time in the conservatorium's history, all students were required to learn a second instrument. Thus, when Piper elected timpani as his second instrument, he began lessons with Alard Maling, who was principal timpanist of the Australian Elizabethan Theatre Trust orchestra and already teaching timpani to Clive Pascoe and Richard Miller. Piper describes Maling as a 'wonderful musician, a great timpanist' (Piper, 2013), but notes that in the 1960s you could not study percussion with Maling or anyone else in Sydney: only timpani lessons were possible. For Piper and his peers such as Barry Quinn in Melbourne, percussion technique was something that was learnt 'on the job' in the Conservatorium Orchestra, the National Training Orchestra (founded in 1967) and the SSO, which Piper joined in 1968. Piper acknowledges that he is 'probably the last percussionist to get a job in one of the symphony orchestras that didn't major as a percussionist' (Piper, 2013) and reflects that his SSO colleagues Lou Tutschka and Maling were valuable mentors early in a career that saw Piper retire as one of the longest serving members of the SSO in December 2013 (Piper, 2012). Although it originally stemmed from a lack of access to formal institutional training in orchestral or contemporary percussion, the establishment of informal mentorships within percussion sections provided the education necessary for young percussionists. There was no contemporary percussion activity in Sydney during the 1960s, and the first concert by The Sydney Percussion was the first percussion ensemble performance experience for all members of the ensemble except Askill. Thus, contemporary percussion technique was developed by Sydney percussionists in the 1970s 'on the job' in percussion or chamber ensembles.

Like Askill, Ron Reeves was a student of Richard Smith's in Adelaide in the early 1970s. Reeves studied with Smith for two years, and in 1973 at the age of 19, moved to Sydney to pursue a freelance career in drum set and percussion. As Piper had before him, Reeves spent a few months in the National Training Orchestra, before starting to work as a casual percussionist with the SSO and AETT. In these ensembles he met Askill, who invited him to be involved in The Sydney Percussion.

The Sydney Percussion's first performance was on 10 August 1974 at the Sydney Opera House Recording Hall, and excerpts of the concert program are shown in Figure 3.1. Curated by Ahern, The Sydney Percussion premièred the percussion ensemble version of Richard Meale's *Interiors/Exteriors* (1970) and Maurice Ohana's *Études Chorégraphiqes*, which was one of the works in Les Percussions de Strasbourg's repertoire. The concert was promoted and reviewed in various newspapers. *Sydney Morning Herald* critic Patricia Brown reviewed the concert positively, without hiding her surprise at the success of a contemporary percussion concert. Brown's statement that 'totally enjoyable and sparkling experiences of contemporary music were still a rather rare phenomenon' (Brown, 1974) gives an indication of the response many Australians had to contemporary music at the time.

Although the group's first performance was both successful and timely – as it occurred during a prolific creative period for the arts in Australia, supported by Gough Whitlam's Labor government – the ensemble did not immediately begin to perform together regularly. A few months after the ensemble's debut, Askill won the principal timpani chair with the Melbourne Symphony Orchestra and left Sydney in early 1975. Over the next couple of years, the remaining members performed only occasional concerts when brought together for chamber music events in Ahern's AZ Music series. However, when Askill returned to Sydney in 1976 as principal percussionist of the Sydney Symphony Orchestra (SSO), The Sydney Percussion began anew. Around this same time, Ron Reeves left Sydney to take up a percussion teaching position in Indonesia. Reeves was replaced by Richard Miller, who had recently returned from a seven-year absence from Australia. During his time abroad, Miller spent five years pursuing a diverse range of percussion interests, studying in North America with Forest Clark and Mitchell Peters of the Los Angeles Philharmonic and in New York with Fred Hinger, who was timpanist with the orchestra of the Metropolitan Opera at that time. An interest in jazz and rock drumming then led Miller to the UK and Europe, where he freelanced for two years before returning to Australia (Sydney Symphony Orchestra, 2017) performing first as a percussionist with the SSO and assuming the position of principal timpanist from 1986 until 2018. When Miller replaced

Reeves, for a time, The Sydney Percussion and the SSO percussion section were one and the same.

For the first 20 years of the ensemble's operation, the SSO provided the group with direct and in-kind support. For example, in the early years of the group's existence, the SSO percussion space became a rehearsal space for The Sydney Percussion, which enabled the group to program large works using various instruments from the orchestra's collection. Later, when the ensemble began to tour internationally, the players were granted leave as the orchestra's General Manager Harold Hort and his successor Mary Vallentine agreed that having a section

Figure 3.1 Excerpts from The Sydney Percussions' first concert program in 1974.

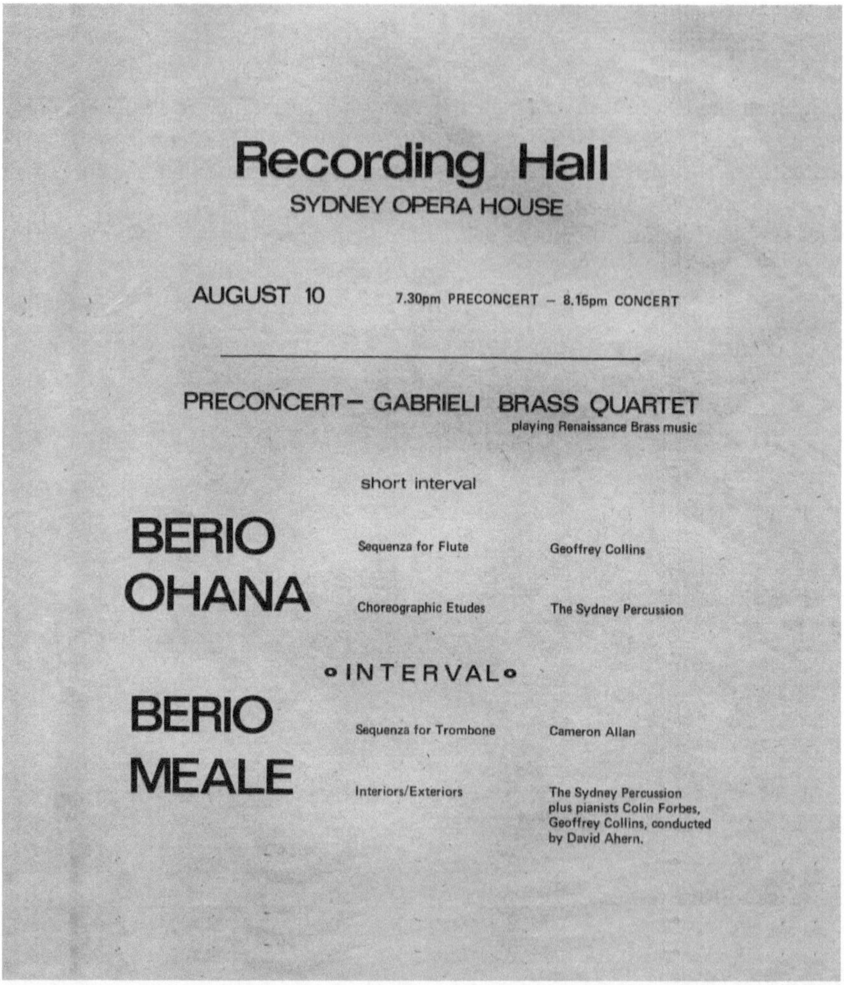

Recording Hall
SYDNEY OPERA HOUSE

AUGUST 10 7.30pm PRECONCERT – 8.15pm CONCERT

PRECONCERT – GABRIELI BRASS QUARTET
playing Renaissance Brass music

short interval

BERIO	Sequenza for Flute	Geoffrey Collins
OHANA	Choreographic Etudes	The Sydney Percussion

○ INTERVAL ○

BERIO	Sequenza for Trombone	Cameron Allan
MEALE	Interiors/Exteriors	The Sydney Percussion plus pianists Colin Forbes, Geoffrey Collins, conducted by David Ahern.

Figure 3.1 (Continued)

with an international profile reflected well on the orchestra. In 1996, the SSO and Synergy collaborated on a CD recording of music by Toru Takemitsu and Carl Vine released by ABC Classics. The SSO acknowledged the importance of the ensemble via formal and informal support, and The Sydney Percussion's first performance with this new membership was presented as part of the SSO Chamber Music Series in 1976.

As a result of the support, The Sydney Percussion began to be involved in more regular performances around Sydney hosted by large organisations such as the ABC. A 1978 ABC concert at the Art Gallery of NSW titled 'Synergy: The Sydney

The Sydney Percussion

Created by Sydney's young leading percussionists for the performance of new music for percussion with or without other instrumental or vocal combinations.

The membership of the SYDNEY PERCUSSION is flexible. Tonight they perform as a group of four alone, and as a group of three in conjunction with two pianists (and conductor) for Richard Meale's Interiors/Exteriors.

Future performances by the group will include Ionization - Varese (6-12 percussionists), Toccatta - Chavez (6 percussionists), Sonata for Two Pianos and Percussion (2 percussionists) - Bartok, Persephasa - Xennakis (6 percussionists), Circles - Berio for voice, harp, and 2 percussionists, How the Stars Were Made - Sculthorpe (6 percussionists), Shen - Tona Scherchen (6-12 percussionists), Sequenza - Berio (3 percussionists), Bombardments - Moran (5 percussionists), Zyklus (1 percussionist) – Stockhausen

The group: Ron Reeves - Percussionist, Elizabethan Trust Orchestra.
Colin Piper - Associate Timpanist, Percussionist, Sydney Symphony Orchestra.
Ian Bloxom - Freelance Studio Drummer/Percussionist, Percussionist, Sydney Symphony Orchestra.
Michael Askill - Freelance Percussionist, Sydney Symphony Orchestra.

Figure 3.1 (Continued)

Percussions' foreshadowed change. Between 1974 and 1979 the names Sydney Percussions, The Sydney Percussions, Sydney Percussion and The Sydney Percussion were used interchangeably. In 1979, The Sydney Percussion changed its name to Synergy Percussion, a title that reflected the group's interests in collaboration around percussion music. In another 1979 program Synergy was defined as a 'combined action; co-operation', and this referred to the collective efforts by the players, composers and their support networks in mounting Synergy concerts. As the group developed, so too did the meaning behind the name Synergy and it was applied to a number of performance situations. For example, the name Synergy reflected the interdisciplinary interests of the group clearly in the 1992 dance and percussion project with Graeme Murphy, *Synergy with Synergy*.

As with a number of percussion ensembles around the world before them, the genesis of Synergy Percussion was the result of efforts by both percussionists and composers. Until 1984, all of Synergy Percussion's concerts were the result of invitations to perform for events run by composers such as Ahern and Alan Holley, pianist Roger Woodward or organisations such as the ABC, the SSO and the Seymour Group.

With the support of external organisations and individuals, the members of Synergy Percussion were able to prioritise the music they were performing above the administration of the ensemble. Many player-run chamber ensembles have a lifespan of less than a decade, which can be attributed to a number of factors. The workload required of musicians who balance a freelance career with the musical, financial and administrative demands of running an ensemble frequently leads to 'burnout', and subsequent dissolution of ensembles. That Synergy Percussion was able to limit the diversions from the music that logistical, organisational, promotional and administrative tasks demand, it can be assumed that the support of external organisations and individuals has been a contributing factor to the group's longevity from the outset. Additionally, and similarly to every Australian percussion ensemble established prior to 2000, Synergy Percussion was led by players who also held other orchestral or freelance positions, thus was founded as a part-time ensemble which was not expected to profit financially. It can be assumed that, had the individual members not held full-time jobs in music, the ensemble would have more keenly felt the financial challenges that arose in the production of independent contemporary music concerts, in addition to logistical concerns. Thus, the support of the orchestra beyond provision of rehearsal space must be acknowledged in the stability provided by the full-time orchestral position held by each Synergy member. The members of Synergy Percussion were quick to pay this goodwill forward, and where the members of Synergy had the most direct impact on student percussionists and colleagues was in Sydney, including at the New South Wales State Conservatorium of Music, now known as the Sydney Conservatorium of Music.

The Sydney Conservatorium of Music

Despite the prominence of professional percussionists in Sydney, tertiary percussion study was not available in New South Wales until 1981, 17 years after percussionists were first admitted to the Elder Conservatorium in Adelaide. Prior to this, a strong secondary school education system had enabled a handful of percussionists to make their way into the profession in Sydney, including Australian Opera and Ballet Orchestra timpanist David Clarence and freelancers Brian Nixon and Graeme Leak. In 1981, Sydney Conservatorium of Music engaged Jack Purdon to teach two students, Tim White and Kim Taylor. In the twilight of his career at the time, Purdon was a percussionist with the SSO who had begun his career in the pit orchestras accompanying silent movies. Frequently requiring percussionists to use unconventional combinations of percussion instruments and everyday objects to create sound effects, the silent movie industry provided a departure from traditional orchestral work in the early twentieth century long before awareness of contemporary percussion or existence of percussion ensembles became part of the Australian percussionist's consciousness. Unfortunately, any link between percussion techniques explored in silent movie work and contemporary percussion in Australia is tenuous at best. With the advent of sound in cinema in the 1920s, sustained work producing live soundtracks for cinema declined decades before Australian percussionists began to explore contemporary percussion techniques and percussionists who had worked in theatres could pass on their knowledge in a percussion ensemble context. Decades later, some theatre percussionists, including Purdon, were among the first generation of educators at the tertiary level; however, they were primarily engaged to teach orchestral percussion. Tim White remembers Purdon talking about his work in the silent movies, but notes Purdon's teaching style did not connect the unconventional percussion instruments of the silent film orchestras with modern percussion ideas:

> He was a link right back to the silent movie days when all the picture theatre orchestras hired buckets of percussionists. Every picture theatre! If it was a cheaper one it had a pianist, but if it was a posh one it had an orchestra, even if it [the orchestra] was just six people it needed a percussionist to do the sound effects. Jack talked about that quite a lot and I loved the idea of having a link back to the silent movies, but he knew nothing about modern percussion.
>
> (White, 2012)

In 1982, Synergy member Richard Miller began teaching percussion at the Sydney Conservatorium, the same year that students Tim White and Peter Jacob, together with composer Michael Whittaker and pianist Michael Kieran Harvey, formed their own, unnamed percussion ensemble. Inspired by Synergy and confronted with limited repertoire available in the library for the new quartet to perform, they began rehearsing some of Synergy's repertoire. White remembers:

> When he heard we were trying to play Synergy's music, Colin Piper very kindly came in and tutored us for nothing whenever we got near to a performance. The level of support, friendship and assistance from all the percussion guys, the players in the Sydney Symphony, Synergy and Brian Nixon, Graeme Leak and David Clarence, they were so helpful. It was really wonderful.
>
> (White, 2012)

In 1985, Daryl Pratt migrated from California to take up a percussion lecturing position at the Sydney Conservatorium of Music, an appointment that signalled further growth in percussion in Sydney. With a background in contemporary music and jazz, Pratt began performing with a range of groups in Sydney and composing new repertoire featuring percussion including vibraphone solo *Webb Spinner* (1986), duo *Sock Bop-a-Dop* (1989) and three works for Synergy titled *Fantasy* (1993), *On the Wings of a Kiss* (1994) and *Alchemy* (1995). Pratt was also active in a number of other ensembles featuring percussion, the most prominent of which were the Australian Art Orchestra and the duo Pipeline with trombonist Simone de Haan. A key feature of Pipeline's work was the performance of Australian repertoire through collaboration with composers and improvisers. Notable was the performance of repertoire by members of the ensemble, which led to the blurring of boundaries between composer, performer and improviser. For example, their 1996 album *In the pipeline* was composed entirely of works by Pratt, de Haan and their collaborator Phil Treloar. Pratt also founded the Sydney Conservatorium Percussion Ensemble in 1991, and since this time has composed a range of repertoire for the ensemble to perform (Pratt & Miller, 2016).

The sense of community and support referred to by White and other percussionists, including Piper, Cleworth, Lagos, Pye and others, was common in orchestras, percussion ensembles and music departments around the country. Although each department was forming independently, some contact between players was possible due to the Australian Youth Orchestra (AYO) National Music Camps, who engaged tutors such as Piper in Sydney and Ryszard Pusz in Adelaide. This contact proved equally important as institutional training, as highlighted by Pye, who cites the AYO as increasing his awareness of percussion in Australia (Pye, 2012). As the number of percussionists active in the Sydney music scene grew, so too did the impact of Synergy Percussion. The individual and collective contributions to Australian music made by members of Synergy Percussion resonated with performers around the country as their programming matured in the 1980s.

Synergy Percussion in the 1980s

The 1980s marked the start of a new period of activity for Synergy Percussion and a number of significant events took place that indicate the ensemble and its members became enmeshed within Australia's cultural fabric during this time. These include securing annual funding, personnel changes, international touring, the cultivation of new approaches to making work, performances of major repertoire and collaborations. Major operational changes took place when the ensemble began to receive regular funding from the Australia Council. This funding allowed the ensemble to invest their energy into creative concerns in self-managed concerts. Later, funding enabled the furnishing of their first studio in Ultimo, and the establishment of the first Synergy Percussion board in 1987. The same year brought a significant personnel change when Miller and Leak left the ensemble to pursue other interests. The original link with the SSO was maintained when they were replaced by Rebecca Lagos and Ian Cleworth, both of whom had taken up positions within the SSO percussion section.

Lagos (b. 1961) grew up in Canberra and, like Piper, came to percussion via the piano. Participation in youth orchestras and national music camps towards the end of secondary school brought Lagos into contact with Miller and Askill, and in

1979 she had a handful of lessons with each of them. Lagos' American godfather had gifted her a subscription to the Percussive Arts Society and this propelled Lagos to commence a Bachelor of Music with George Gaber at Indiana University, Bloomington, in 1980. Although Gaber had visited Australia just years earlier, Lagos was unaware of his visit at the time and her choice to study with Gaber was coincidental. Despite the fact that the sheer size of the music school at Indiana University Bloomington was greater than anything that existed in Australia at the time, Lagos notes that her experiences with Synergy founding members before she left had a strong impact. When she returned to Australia in 1984, Lagos spent three years as Tasmanian Symphony Orchestra (TSO) principal timpanist, then joined Synergy in 1987, the same year she won a position with the SSO.

Cleworth has been mentioned previously as another student of Richard Smith's in Adelaide; training that led to principal positions in the Queensland Orchestra (1983–1985), the Sydney Symphony (1986–2006) and ensemble Adelaide Percussions. Cleworth is discussed in Chapter Four. Cleworth and Lagos brought a new energy to the group and the quartet of Askill, Piper, Cleworth and Lagos would remain intact for ten years. Askill remained a key figure within the group's activities and his leadership has been acknowledged by his colleagues as pivotal in the group's development. Lagos cites Askill as the catalyst for her percussion career and Cleworth describes Askill as 'an incredible conceptual thinker and an amazing creative brain. Michael's leadership inspired me' (Cleworth, 2013). Figure 3.2 shows Cleworth, Askill, Lagos and Piper with part of their instrument collection at the first Synergy rehearsal space in Ultimo, Sydney. Piper and Lagos both cite the acquisition of this rehearsal space as a significant step in the ensemble's career.

In addition to operational changes, it was in the 1980s that the ensemble began introducing major international repertoire by composers such as Iannis Xenakis, John Cage and Steve Reich to Australian audiences. These works highlight the broadening scope of the group and the increasing level of technical proficiency achieved by the players. In reflecting on his predecessors in Synergy, Cleworth notes that Askill, Leak, Miller and Piper were 'musically, technically and conceptually raising the bar' and that this 'had an influence on players all around the country, obviously me included' (Cleworth, 2013). A student in Canberra at the time, Rebecca Lagos remembers 'I was very heavily influenced by Michael and Rick I think and the way they played, their flair and their sort of musicality' (Lagos, 2013). Despite the commitment of the ensemble members and a greater level of external support than previously held, not everything was easily achieved. Sourcing instruments and materials for increasingly more ambitious projects was challenging, and not infrequently, last minute changes had to be made to programs. For example, in 1989, Synergy Percussion used the *Sydney Morning Herald*'s 'Column 8' to involve the wider community in the gathering of objects such as four cacti for the performance of John Cage's *Branches* (1976), shown in Figure 3.3. However, Piper's meticulous records show that no performance of *Branches* took place in 1989 or in the subsequent years. Similarly, in 1985, upon arriving at a concert venue and discovering no piano, Cage's *Credo in US* (1942) was necessarily cut from the program at the eleventh hour.

Figure 3.2 Ian Cleworth, Michael Askill, Rebecca Lagos and Colin Piper at the first Synergy Percussion rehearsal space in Ultimo, Sydney.

Source: Photograph by Daniel Mendelow.

Of greatest significance was Synergy Percussion's newly concerted focus on accelerating the development of percussion practices in Australia through performance and recording. The style of programming developed during the 1980s that presented commissions from Australian composers alongside major international repertoire is maintained by the group today. In 1981 they presented their first entirely Australian program, featuring music from three generations of Sydney-based composers. As part of a series organised by contemporary chamber ensemble The Seymour Group, Synergy Percussion performed their first commissions: *Percussion Quartet* (1980) by Colin Bright and *Hydra* (1980) by Gerard Brophy, alongside Peggy Glanville-Hicks' *Sonata for Piano and Five Percussion* (1951) and Peter Sculthorpe's *How the Stars Were Made* (1971). This marked one of, if not the first formal performance of Peggy Glanville-Hicks' music since her return to Australia in 1975. This concert marked a conscious decision by Synergy to place a strong emphasis on the programming of Australian works, thereby signalling their encouragement of Australian composers to write for the genre. Due to the fact that the Australian concert-going public was not yet acclimatised to a practice that would mature into one of the ensemble's strengths, the concert received

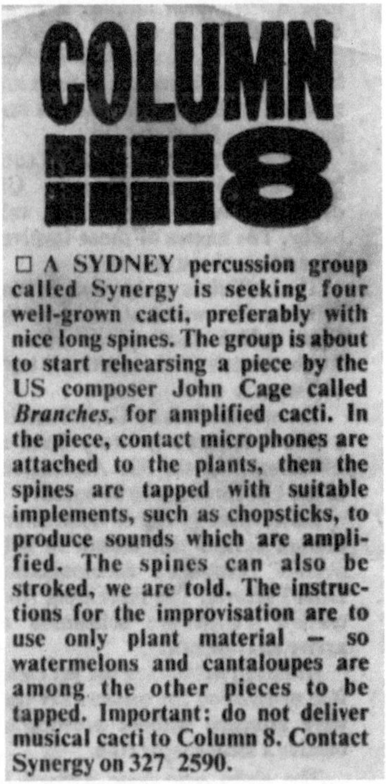

COLUMN 8

☐ A SYDNEY percussion group called Synergy is seeking four well-grown cacti, preferably with nice long spines. The group is about to start rehearsing a piece by the US composer John Cage called *Branches*, for amplified cacti. In the piece, contact microphones are attached to the plants, then the spines are tapped with suitable implements, such as chopsticks, to produce sounds which are amplified. The spines can also be stroked, we are told. The instructions for the improvisation are to use only plant material – so watermelons and cantaloupes are among the other pieces to be tapped. Important: do not deliver musical cacti to Column 8. Contact Synergy on 327 2590.

Figure 3.3 A *Sydney Morning Herald* 'Column 8' notice placed by Synergy seeking four cacti for the proposed 1989 performance of John Cage's *Branches*.

mixed reviews. Despite some bewildered comments from *Sydney Morning Herald* reviewer Fred Blanks that 'in an all-percussion concert [composers] must find it hard to think of something new' (Blanks, 1996), Synergy was not dissuaded from performing local repertoire. Indeed, the ensemble developed a reputation for supporting the work of emerging and established Australian composers through various commissioning projects that were made possible by Australia Council funding as well as industry sponsorship by organisations such as 2MBS radio (Martin, 1996). The performance of new Australian repertoire became a major area of development and as Synergy Percussion continued to produce high quality performances of new music, the reputation of and activity around the ensemble grew.

A number of significant recordings were made during this period, including *Australian Percussion* (Askill, 1986) and the 1989 *Anthology of Australian Music CSM:2* compact disc, multi-tracked in its entirety by Askill, who remembers:

> I decided to record and compile some important percussion pieces by composers that I was working with closely at that time: Nigel Westlake, Peter Sculthorpe, Ross Edwards and Martin Wesley-Smith. The pieces were recorded on the first digital multi-track (by Sony) to appear in Sydney at a studio near the Sydney Town Hall. *Omphalo Centric Lecture* was recorded at the old ABC

> studios near Kings Cross and produced by Nigel Westlake. Though it was not our original intention to multi-track the piece but rather to record a Synergy version, it turned out that the other players were unavailable due to Sydney Symphony commitments. So the multi-tracking idea was discussed and the challenge accepted!
>
> (Askill, 2017)

A wave of new Australian percussion repertoire

Prior to 2000, Australian music critics often referred to the 1980s as a golden age for Australian percussion. This was prompted by the sudden explosion of percussion performances and the percussion repertoire that was being created in Australian capital cities at the time by Synergy and the various ensembles discussed in subsequent chapters. From the outset, Synergy championed a range of contemporary percussion musics including notated and semi-improvised works, and the ensemble presented international and Australian premières at each concert for many years. As a result of his study in Strasbourg, Askill introduced some established percussion ensemble repertoire; however much of the group's early repertoire consisted of works composed and suggested by local colleagues. As the Australian Percussion Ensemble had done before them in Melbourne, this set the scene for a strong collaborative practice amongst percussionists in Sydney. This approach was similarly favoured by ensembles later established in other cities. A commonality between the output of all of these groups can be found in the effective, long-term relationships established between composers and performers that led to a diverse and numerous catalogue of new works.

The Australian repertoire composed for Synergy in the 1980s and into the 1990s helped the group to form a unique identity. This was influenced by three factors. First, the freedom granted to composers enabled a wide variety of repertoire to be created. Second, Synergy's ties with the SSO gave them access to instruments unusual in Australia at the time that gave the ensemble a unique sound. The instrumentation of any given percussion commission is dependent on the performer's instrumentarium, and the access to the SSO studio significantly diversified Synergy's sound. For example, the SSO purchased a set of chromatic almglocken (Swiss cowbells) in the preparation for a performance of Olivier Messiaen's *Et Exspecto Resurrectionem Mortuorum* in the 1970s that Synergy were able to use in other programs. Third, it provided an ongoing opportunity for the members of Synergy to collaborate with composers, explore a range of ideas and ultimately to experiment with their own compositions.

Thus, collaborations between composers and Sydney percussionists and percussionist-composers became more frequent. For example, in 1983, Touchwood was formed, a trio comprising Askill and composer-clarinettist Nigel Westlake and Graeme Leak (then based in Melbourne). Although Touchwood was active for only a few years, they generated a relatively high volume of new repertoire, as each performer also composed for the trio. The majority of this repertoire was for two marimbas and bass clarinet, however a number of percussion-only versions of Touchwood repertoire were later created. For example, the original version of Westlake's marimba quartet *Omphalo Centric Lecture* (1984) was for the instrumentation of the Touchwood trio. Westlake acknowledges Touchwood's role in the development of *Omphalo Centric Lecture*:

> It started life as a piece for two percussionists and bass clarinet. I was playing [bass clarinet] in a trio called Touchwood with percussionists Michael Askill and Graeme Leak. Michael suggested I re-work *Omphalo* for Synergy . . . It soon became a regular part of their programs and Michael was the first player to record it – as a multi-tracked performance.
>
> (Westlake, 2005)[1]

In the mid-1980s Askill left the Sydney Symphony Orchestra and his increasing work with smaller ensembles including Michael Atherton's Table Band with cellist John Napier and percussionist Jess Ciampa (later renamed Southern Crossings) and creative collaborations with Nigel Westlake. Askill and Westlake's duo The Cast of Thousands led to numerous small projects and development of works for concert performance and film soundtracks. These projects were also the stimulus to begin composing works combining acoustic percussion and electronics such as the Mallet Kat and 'software developed for the Atari computer as well as the emerging range of synthesizer sequencers incorporating samplers by Roland, Yamaha, Ensoniq and Akai' (Askill, 2013b). This technology also aided in the composition of acoustic works. Askill's first concert piece was composed for Synergy in 1989. Titled *No Trace*, this work for percussion ensemble with solo flute was composed using a multi-track synthesizer sequencer. This was followed by *Lemurian Dances* (1990) for percussion quartet and tape. Askill reflects on *Lemurian Dances*:

> Composing music for mallet instruments is not something I find easy. For *Lemurian Dances*, I decided to improvise the mallet parts and notate the notes afterwards. I did this using a Mallet Kat – one of the first midi controllers designed for percussionist. I notated the improvisations exactly as they were received via the Kat and this included a technical glitch inherent in the Kat at that time – pads would sometimes misread the velocity of a stroke or sometimes not trigger at all. This resulted in some unintended accents and syncopations that were included in the final notation.
>
> (Askill, 2017)

It was during this period that Synergy also placed importance on the regular performance of works composed by members within the group. Many of these are now amongst the most well-known of Australian percussion ensemble works. Askill reflects that Synergy became his first serious compositional vehicle because he 'felt comfortable with the instruments and with the players and felt ready to test more ideas' (Askill, 2013b). This sentiment has been echoed by other Synergy players, including Ian Cleworth, Graeme Leak, Timothy Constable and Bree van Reyk.

In addition to dozens of international works, Synergy Percussion has premièred and/or commissioned over 130 works from Australian composers to date. Compiled from various archival collections and interviews with percussionists and composers affiliated with Synergy, Table 3.1 provides a full list of works commissioned between the ensemble's formation in 1974 and 2000, encapsulating the diverse spectrum of percussion repertoire performed by the group. It also shows that the number of new Australian works performed by the group steadily increased during their second decade of operation. Synergy Percussion had commissioned 13 Australian works by 1984, and by 1994 this had grown to a total of 61 commissioned works. The rate of new works continued to grow at an inexorable rate, in part due to the fact that the performance of works composed or arranged within the group became a key part of Synergy's creative practice. It is pertinent to note that there are a number of works made popular by the group that were not commissions, and are therefore not included in the Table 3.1. These include works that Synergy Percussion premièred but did not commission, such as Michael Smetanin's *Speed of Sound* (1983), James Ledger's arrangement of Ross Edwards' *Dance Mantras for Brass and Percussion* (1992) or Ross Edwards' *Flower Songs* (1987), and a number of solo works that were written for individual Synergy members that were performed at Synergy Percussion concerts, such as the solo work *Percussion Music One* (1984), composed for Graeme Leak by Phil Treloar.

Further, through the 1980s and 1990s, composers from around Australia and New Zealand sent unsolicited compositions to Synergy, many of which were subsequently programmed by the ensemble. Synergy readily embraced new works by composers such as Carl Vine (b. 1954), Colin Bright (b. 1949), Nigel Westlake (b. 1958), Ross Edwards (b. 1943) and Robert Irving (b. 1943), with whom they formed long-term professional relationships. These relationships yielded a number of cornerstone Australian percussion works, including *Omphalo Centric Lecture* (1984), *Defying Gravity* (1987) and *Prelude and Dragonfly Dance* (1991). According to Askill, the foundations for the formation of long-term relationships with composers were laid during his percussion training in youth orchestras, and Askill credits John Hopkins with making collaboration between performers and composers a common part of his artistic practice throughout his career. Hopkins' regular inclusion of music by emerging and established Australian composers during various Australian Youth Orchestra (AYO) programs provided introductions that would prove integral to the making of new work in the future and ensured that interaction between performers and composers was a natural part of music training at the time. Askill confirms that the experience of working with composers became common when he states: 'You'd be working up the back with Peter Sculthorpe or Barry Conyngham or Richard Meale looking over your shoulder. It was normal that you'd be working and talking and communicating with living composers' (Askill, 2013a).

As the Australian Percussion Ensemble had done before them, members of Synergy explored improvisational works as well as works with tape or electronics. These include Martin-Wesley-Smith's *For Marimba and Tape* (1982), *Silençio* (1987) for percussion, live electronics and graphics, and Carl Vine's *Percussion Concerto* (1987) for percussion solo and tape. Featuring a diverse blend

Table 3.1 Australian repertoire commissioned by Synergy Percussion between their formation in 1974 and 2000

Year	*Composer*	*Work*	*Instrumentation*
1977	Robert Irving	*Gong Song*	Trio (optional doubling)
1980	Gerard Brophy	*Hydra*	Quartet
1980	Colin Bright	*Quartet*	Quartet
1970	Richard Meale	*Interiors/Exteriors*	Quintet (chamber version)
1982	Trevor Pearce	*Deserts I*	Quartet
1982	Martin Wesley-Smith	*For Marimba and Tape*	Marimba solo
1984	Stephen Loader	*Piece for Percussion*	Quartet
1984	Benjamin Thorn	*Esparto Grass*	Quartet (workshop piece)
1984	Jonathan Mustard	*4 Pieces for Percussion and Tape*	Quartet
1984	Matthew Jones	*Moments of Repose*	Quartet (workshop piece)
1984	Stewart Eagles	*7 : 5 : 3*	Quartet (workshop piece)
1984	Stephen Loader	*Piece for Percussion*	Quartet (workshop piece)
1984	Colin Bright	*Tulpi Stick Talk*	Quartet
1985	Ross Edwards	*Reflections*	Quartet
1985	Julian Yu	*Scintillations II*	Quartet
c. 1985	Martin Wesley-Smith	*White Knight and Beaver*	Arranged for marimba and xylophone[2]
1985	Gillian Whitehead	*Windstream*	Percussion solo
1985	Gillian Whitehead	*Tongues Swords Keys*	Percussion and choir
1986	Nigel Westlake	*Le Cirque de Demain*	Quartet
1986	Mark Isaacs	*Memoirs*	Percussion trio and piano
1986	Joan Brassil and Synergy Percussion	*Can it be That Everlasting is Everchanging*	Improvisation
1986–87	Martin Wesley-Smith	*Silencio*	Quartet
1987	Carl Vine	*Defying Gravity*	Quartet
1987	Bozidar Kos	*Quasar*	Quartet
1987	Carl Vine	*Percussion Concerto*	Solo with tape
1988	Stephen Bull	*Ball the Jack*	Marimba and tape
1988	Michael Askill	*Procession*	Quartet
1989	Synergy Percussion	*Green Piece*	Quartet
1989	Gillian Whitehead	*Napier's Bones*	Piano and 24 improvising percussionists
1989	Andrew Schultz	*Machine*	Quartet
1989	Nigel Westlake	*Moving Air*	Quartet and tape
1989	Graeme Leak	*Rubadub*	Trio
1989	Nigel Westlake	*Happy Birthday Tribute*	Quartet
1989	Martin Wesley-Smith	*Happy Birthday Tribute*	Quartet
1989	Ross Edwards	*Happy Birthday Tribute*	Quartet
1989	Michael Askill	*No Trace*	Quartet
1990	Michael Barkl	*Disco*	Quartet and sequencer
1990	Stephen Bull	*Thieves*	Quartet
1990	Michael Askill	*Lemurian Dances*	Quartet
1990	Nigel Westlake	*Malachite Glass*	Percussion quartet and bass clarinet
c. 1990	Philip Treloar	*New Forms 3*	Quartet[3]
1991	Ross Edwards	*Prelude and Dragonfly Dance*	Quartet

(*Continued*)

Table 3.1 (Continued)

Year	*Composer*	*Work*	*Instrumentation*
1991	Andrew Ford	*Alchemy*	Quartet
1991	Sharon Calcraft	*Meditation d'un peintre*	Conductor, solo voice and quartet
1992	Julian Yu	*Quartet*	Quartet
1992	Robert Douglas	*Arcs and Heresies*	Quartet
1992	Claire Jordan	*Zed*	Quartet (for 2MBS composer competition)
1992	Raffaele Marcellino	*Plusperfect Squaredance*	Quartet (for 2MBS composer competition)
1992	Gerard Brophy	*Glint*	Sextet
1992	Michael Askill	*The Wave*	Taiko quartet (for *Matsuri* project)
1992	Michael Askill	*A Clear Midnight*	Quartet (for *Matsuri* project)
1992	Michael Askill	*Going On*	Quartet (for *Matsuri* project)
1993	Mike Nock	*Time-lines*	Improvising soloist and percussion ensemble
1993	Daryl Pratt	*Fantasy*	Quartet
1993	David Lumsdaine	*Rain Drums*	Quartet
1994	Stephen Cronin	*KISS*	Mallet quartet
1994	Daryl Pratt	*On the Wings of a Kiss*	Percussion quintet and jazz ensemble
1994	Cathie Travers	*Weird Dreams*	Quartet and sampler
1994	Ian Cleworth	*Hour of the Dragon*	Percussion quartet
1994	Ian Cleworth	*Takeda's Poem*	Percussion quartet
1994	Ian Cleworth	*Go*	Percussion quartet
c. 1995	Ian Cleworth	*Taiko Fanfare*	Quartet
1995	Andrew Schultz	*Chorale, Demon, Beacon*	Percussion quartet and bass koto
1995	Gerard Brophy	*Trash*	Junk percussion quartet
1995	Daryl Pratt	*Alchemy*	Quartet
1996	Gerard Brophy	*Umbigada, obrigado*	Percussion quartet
1996	Michael Atherton	*Exhortation*	Percussion quartet and SATB vocal ensemble
1996	Michael Askill	*Free Radicals Suite*	Trio
1996	Nigel Westlake	*The Invisible Men*	Quartet
1996	Colin Bright	*The Wild Boys*	Percussion quartet and saxophone quartet
c.1996	Ian Cleworth	*Moon Over Water*	Quartet
1996	Ian Cleworth	*Matsuri Daiko*	Quartet
1996	Ian Cleworth	*Winter Haiku*	Percussion Quartet
1996	Ian Cleworth	*Summer Haiku*	Percussion Quartet
c. 1998	David Hewitt	*Autumn Haiku*	Japanese speaker and percussion sextet
c. 1998	David Hewitt	*Spring Haiku*	Japanese speaker and percussion sextet
1998	Elena Kats-Chernin	*Umcha Piece*	Quintet
1998	Liza Lim	*Anactoria*	Sextet
1999	Synergy Percussion	*Zoon*	Improvisation-based sextet

of improvised and notated works with an emphasis on music created for the performers that reflected their personal musical interests began to emerge as a model for Australian percussion ensembles. The ability of an ensemble to present diverse yet personal programs that displayed wide ranging technical, stylistic and interpretative skills became a common way to run a percussion ensemble in Australia, as shown in the programming of South Australia's Adelaide Percussions and Western Australia's Tetrafide Percussion, examined in Chapters Four and Five.

Reflecting on the body of repertoire produced for Synergy Percussion, it can be observed that different external influences emerged during different periods of activity. Highlighted in Tables 3.1 and 3.2 and discussed in Chapter Seven, these external influences include the European avant-garde and Australia's geographical context in the 1970s, Asian and African traditional music in the 1980s, and an increasing blend of influences from the 1990s to present day including experimental electronics, pop influences and world musics. These influences were then passed on to the wider percussion community as others performed Synergy's repertoire.

The influence of individual artists both within and without the ensemble is also evident. For example, Nigel Westlake performed frequently enough with Synergy Percussion to have had an impact on the repertoire generated for the group. There are a number of bass clarinet solo works that were written for Westlake and premièred in Synergy Percussion concerts, such as Westlake's *Onomatopoeia* (1984) for bass clarinet and digital delay, *Malachite Glass* (1990) and Michael Smetanin's *The Ladder of Escape* (1984) for bass clarinets.

In his survey of Australian contemporary composers, Gordon Kerry stated that 'music needs performers in order to flourish, and new music needs performers with a certain kind of dedication' (Kerry, 2009: 130). This is certainly true of Australian composers of percussion ensemble music, as the majority of Australian works were composed in the way that they were because of the respect the composers had for the performers for whom they were writing. Indeed, it can be assumed that the breadth of musical styles found within Synergy's catalogue of commissions was possible because composers were inspired by the Synergy players to freely explore previously untested ideas with the knowledge that their skilled Synergy colleagues would strive to execute their work as they imagined. In supporting this assumption, Martin concludes that Synergy's influence on shaping the Australian percussion identity was in part due to creative freedom granted composers by the ensemble. Martin wrote, 'the development of Australian percussion music has been shown to be closely linked to the existence and development of the Synergy percussion ensemble' (Martin, 1996: 119) and attributed the wide range of timbral, textural and rhythmic styles produced for the group to this creative freedom.

The influence of performers is not infrequently missing from music history, however the impact of the Synergy percussionists on the development of percussion in Australia is impossible to ignore. When reflecting on his percussion works, composer Vinko Globokar observed that 'without the performer, the work cannot exist' (Globokar, 1992: 78). In this statement Globokar is referring to the personal abilities and nuances each individual performer brings to a realisation of his work in the use of their voice and body as sources of percussive sounds. The approach of the Synergy performers to sound making and percussive technique is acknowledged as significant in the development of their Australian repertoire. In a review of the 1991 CD recording *Synergy Percussion*, Graeme Skinner observes that 'one of the most interesting things about most of the music on this disc is the way it sounds first typical of Synergy, and only later typical of its composer' (Skinner, 1991). The influence of Synergy on Australian music is also evident in other recordings. The Anthology of Australian music on disc, for example, is a celebrated collection of 30 compact discs recorded with the aim to 'advance knowledge of Australian music in general' (Covell, 1989). The second disc in this collection is not an orchestral recording or another more established western art music form accepted by the Australian audiences of the time, but a Synergy Percussion recording of works by Ross Edwards, Carl Vine, Trevor Pearce and Gerard Brophy.

A number of works composed for Synergy became synonymous with the group's identity from their première date. Over time through multiple performances, some of these works became platforms from which their composers improved the trajectory and diversity of their careers. Writing for Synergy in the 1980s was often the first opportunity for composers to work with percussionists, and the 'newness' of such a commission frequently resulted in the composer commencing a previously unexplored facet of their compositional voice. Many works composed for Synergy have become amongst said composers' most frequently performed repertoire. An example of a work that became important for the composer and the ensemble through Synergy's performances can be found in the aforementioned *Omphalo Centric Lecture*, which Kerry defines as Westlake's first significant concert work (Kerry, 2009: 98). Largely responsible for launching Westlake's international career, *Omphalo Centric Lecture* is also one of, if not the most frequently performed Australian percussion ensemble works, and has been performed by Synergy on over 100 occasions. Its popularity not only increased the frequency of international performances of Westlake's work, but also prompted Westlake to establish his own publishing company, 'rimshot music'. The work has also been the subject of numerous scholarly papers, arrangements and recordings (Dalton, 2006; Ice, 2012; Synergy Percussion, 2004). The ongoing role of composer-performer collaboration in the output of Synergy Percussion was acknowledged in Synergy's 40th anniversary project 40under40. This project included 40 miniature works for percussion and video, commissioned from a cross section of contemporary composers, choreographers, musicians and video artists. This included long-term collaborators Larry Sitsky, Gerard Brophy and Marcus Whale; members of the ensemble Timothy Constable and Bree van Reyk; established and emerging Australian composers Cat Hope, Elena Kats-Chernin, Lachlan Skipworth, Julian Day, Amanda Cole and Kate Moore; and international artists Ruben Naeff and Michael Gordon.

Beyond Australian shores

In the late 1980s, Synergy began to perform at major international percussion and chamber music festivals. This allowed the group to discover how their practice balanced with the international percussion practices of the time, while propelling the group into the international contemporary music scene. Prior to the 1989 Bicentennial, the Australian Embassy in Paris contacted the Australia Council Performing Arts Board seeking suitable Australian cultural representatives for the festival. Advertised as 'Australia's gift to the French Bicentennial celebrations', Synergy Percussion was invited to join 260 percussionists from different countries for Tambours '89 in Paris. Synergy's first international festival appearance was also the first time Australian percussion had appeared on an international platform.

Tambours '89 was viewed as an important opportunity that could act as a springboard for an annual international touring program for Synergy and support from Musica Viva, the Australia Council and the Department of Foreign Affairs and Trade was essential as they discovered 'enormous problems existing in the movement of the huge amount of freight' (Ludgate, Cleworth, Askill, & Piper, 1989). Musica Viva was pivotal in the support of the group, organising sponsors to ensure various obstacles that arose for a touring percussion group were overcome, not least the freighting of 25 instrument cases totalling 900 kilograms over 17,000 kilometres.

Synergy performed two programs of Australian contemporary percussion music at Tambours '89 and in an additional concert organised by the Australian Embassy. The significance of this tour was recognised by the wider Australian musical establishment, and Peter Sculthorpe was commissioned to write a work to be premièred during the tour. Scored for vibraphone, two marimbas, tam-tam, thunder sheet and two rain sticks, Sculthorpe's commission is known by two names, the original *From Jabiru Dreaming* and *Sun Song*. The latter title connects this work to the Sun Song series of compositions Sculthorpe began in 1989, a collection of works frequently cited by scholars as 'Australian' in sound (Covell, 1967; Kerry, 2009; Paget, 2013) due to the incorporation of indigenous melodies and soundscapes depicting the landscape. This concept will be explored further in Chapter Seven. Although *From Jabiru Dreaming* uses material from the Sun Song series, Sculthorpe based the piece's main themes on 'three Aboriginal melodies as transcribed by a crew member of the first French fleet to reach Australian shores' (Ludgate et al., 1989: 3) to acknowledge the event for which the work was created. Sculthorpe accompanied Synergy on the tour in order to introduce the première contributing to the positive reception the performance received. This was reported on by ABC broadcaster Christopher Lawrence, who was also in attendance.

During the course of this tour and others, meetings with international colleagues vital to the development of Synergy Percussion took place, exploring artistic opportunities and administrative approaches including a discussion with Les Percussions de Strasbourg on the challenges faced by a touring percussion ensemble as mentioned previously. In 1990 Synergy performed at Huddersfield Contemporary Music Festival. During this tour, Askill, Cleworth and Lagos gave a private performance of Toru Takemitsu's *Rain Tree* for Takemitsu and their accompanying family members. According to Piper, the most pivotal of the meetings made possible by touring was with Iannis Xenakis, where specific details of his works and their interpretation and the possibility of a commission for Synergy were discussed. Although such a commission did not eventuate, the significance of this meeting for Synergy highlights the importance placed on the Australian première of Xenakis' sextet *Persephassa* in December 1984 by past and present members of Synergy (Cleworth, 2013; Lagos, 2013; Piper, 2013). Performance of Xenakis' virtuosic repertoire is considered as something of a rite of passage to musical maturity for percussionists around the globe (see J. H. Beck, 1995; Harley, 2004; Kanach, 2010; Schick, 1995, 2006a). *Persephassa* was programmed alongside Varèse's *Ionisation*, Australian Colin Bright's *Tulpi Stick Talk* and an arrangement of traditional African rhythms by Graeme Leak. The 1984 performance of *Persephassa* is remembered as an artistic turning point for Synergy by the performers, critics and various audience members.

Askill notes that performing in the context of international music festivals and percussion events gave Synergy the opportunity to objectively view their work and to consider their own sound and identity in direct comparison with their peers. For example, the Taipei International Percussion Convention, which Synergy attended in 1996, was the first time an Australian percussion group had been billed with the same profile as other major international percussion acts including Les Percussions de Strasbourg, Kroumata (Sweden), Amadinda (Hungary) and Talujon (USA). Prior to the Taipei convention, the Synergy percussionists had been operating independently, as geography prevented exposure to and collaboration with their European and American counterparts. This isolation enabled the group to forge their own unique style that they presented in Taipei through the prism of an almost entirely Australian program. Every artist at the convention performed a work by a Taiwanese composer in their program. For Synergy Percussion, it was Loong-Hsing Wen's *Double Tones*. With the exception of this work, Synergy presented an Australian program of repertoire by Sydney-based composers: Nigel Westlake's *Omphalo Centric Lecture* (1984), Ross Edwards' *Reflections* (1985), Michael Askill's *Lemurian Dances* (1990), and two 1996 works by Ian Cleworth titled *Moon Over Water* and *Matsuri Daiko*. Thus, no overlap of repertoire occurred between Synergy and their peers.

Synergy therefore presented a snapshot of the Australian percussion identity of the time in Taipei: an identity that contained references to various influences on Australian percussion repertoire. In reflecting on their program, Askill notes the opportunity to present a unique voice this programming provided:

> I think it did stand apart because although the music wasn't incredibly complex, it had strong rhythmic elements and strong ensemble integration. . . . Ross Edwards [*Reflections*] showed that particular delicate aesthetic too, of Australian composers. . . . I think that did present an identity and a sound that perhaps they hadn't heard before, and that was really appreciated.
>
> (Askill, 2013a)

Another event that shaped the future direction of Synergy was the 1998 festival orchestrated by Swedish percussion ensemble Kroumata in celebration of their 20th anniversary. Held in Stockholm, Kroumata invited performances from the most influential percussionists of the time, including NEXUS (Canada), Amadinda (Hungary), Percussion Group Cincinnati (USA), Talujon (USA), Synergy, Fritz Hauser (Switzerland), Keiko Abe (Japan) and Steven Schick (USA). It was during this festival that Synergy inaugurated an ongoing collaborative relationship with Hauser. Centred around a drum set augmented with percussion, Hauser's practice of improvisation and composition in real time focuses on the gradual exploration of timbre and colour rather than rhythmic drive. Prior to experiencing his performance at Kroumata's celebrations in Stockholm, the members of Synergy Percussion had not experienced Hauser's explorative performance style and were struck by his unique musical voice. Thus, they were surprised and honoured by Hauser's invitation to collaborate. In reflecting on this invitation, Cleworth attributes Synergy's distinct style was at the root of Hauser's suggestion: 'Synergy had a particular sound he responded to, we had a sound concept and approach that was unique, and he felt there was an opening there he could work with' (Cleworth, 2013). Hauser's relationship with Michael Askill and with the current Synergy members has flourished since this time, and his influence in Australia can be found in collaborations with other Australian percussionists including Speak Percussion in Melbourne, Vanessa Tomlinson in Queensland and Gary France in Canberra.

Interdisciplinary collaboration

It was during the 1990s that collaborations with significant contributors to the international contemporary percussion scene became a regular part of Synergy's programming. Since this time Synergy Percussion has regularly collaborated with a diverse group of jazz, world and classical musicians, including Trilok Gurtu, Dave Samuels, Evelyn Glennie, Hossam Ramzy, Noreum Machi and Omar Faruk Tekbilek among others both overseas and in Australia.

The establishment of collaborative relationships was an important aspect of Synergy's artistic practice, and the practices of individual Synergy members. Askill states that a growing awareness of the arts in Australia and internationally, combined with general personal growth and change of direction, stimulated the group's desire to find ways to collaborate artistically with performers from different musical disciplines to create new works:

> African drummers or Japanese koto players and Japanese drumming or music, or traditional jazz players like Phil Marrick or David Jones, or a Middle Eastern musician . . . we could somehow bring contemporary percussion together with those sorts of traditions and gradually build our own kind of tradition through interacting with long standing traditions.
>
> (Askill, 2013a)

Some of the projects for which Synergy became widely known in Australia were interdisciplinary collaborations with Australian artists such as choreographer Meryl Tankard, didgeridoo soloist William Barton and shakuhachi grand master Riley Lee. In 1991 Synergy joined together with Riley Lee, Satsuki Odamura and a *butoh* dancer, Chin Kham Yoke to create the collaborative work *Matsuri*. This work was Synergy's first large-scale attempt at melding contemporary percussion and Japanese traditions into a unified programmatic work. It became the catalyst for future collaboration with Lee that eventually led to the formation of Japanese-Australian contemporary taiko group Taikoz, and propelled Australian choreographer Graeme Murphy to approach Askill to suggest they work together in 1992.

Murphy and Askill had previously met in Strasbourg in the early 1970s, when Murphy was working with dance company Ballets Felix Blaska. Some of the work Murphy was involved in with this company featured performances with live music provided by percussionists. The rhythmic nature of contemporary dance and the visual aspects of contemporary percussion enable the boundaries between the two art forms to be explored cohesively, and Askill and Murphy have since produced a number of collaborations that explore the nexus where contemporary dance and contemporary percussion meet. The first of these was *Synergy with Synergy*, a work that featured the Sydney Dance Company performing new choreography by Murphy to Synergy's live performance of existing works by Nigel Westlake, Ross Edwards, Michael Askill, István Márta, Elliot Carter and John Cage. Askill reflects on the impact the collaborations with Murphy had on his and Synergy's practice:

> I think that connection with dance was a big thing because it seemed to consolidate the role of percussion, the role of music as an interactive and integral part of dance and that was quite exciting.
>
> (Askill, 2013a)

Cited as 'one of the most remarkable collaborations in the world of dance theatre' (Schaefer, 2007), the significance of Askill's collaboration with Graeme Murphy and the Sydney Dance Company is repeatedly highlighted in the media: '*Synergy with Synergy*, the critically acclaimed 1992 collaboration with Graeme Murphy's Sydney Dance Company brought Synergy into direct contact with a whole new artistic world – dance and theatre – and a new audience' (Synergy Percussion, n.d.). The new audiences to whom the preceding quote refers were both national and international. Sydney Dance Company and Synergy later toured this production to Brisbane, Taipei and Tokyo. Following this, a second production was created in 1996 for three percussionists called *Free Radicals*. Created with touring in mind, *Free Radicals* went further than *Synergy with Synergy* in the blurring of distinction between dancer and percussionist as the percussionists engaged in dance-like performance and vice versa. Askill's music direction in this work included musical directives that involved the dancers. For example, among the instruments available to the musicians were the dancers' bodies, and rhythmic patterns played by the percussionists were spoken by the dancers in various languages to add a dimension to the music. In subsequent collaborations between Askill and Murphy, including *Salome* (1998), *Party* (1998), *Air and Other Invisible Forces* (1999) and *Hua Mulan* (2005), interdisciplinary interaction continued to be a focus, taken to still more interconnected levels particularly in *Salome*, which also involved Synergy Percussion performers.

Askill's interest in collaboration was shared by the other members of Synergy; however, much of the early collaborative relationships and processes were driven by Askill. These include projects with world musicians Omar Faruk Tekbilek and Hossam Ramzy. Projects that focussed on bringing Japanese taiko elements together with contemporary percussion were, however, facilitated by Cleworth. Additionally, Askill and Cleworth engaged independently in numerous collaborations with other artists that they continued when Cleworth and Askill left Synergy in 2007 and 2010, respectively.

Forty years of Synergy Percussion

In 2014, Synergy Percussion celebrated 40 years of contributions to Australian contemporary music. The history of the group is maintained via ongoing collaborative relationships with composers and performers who were part of the early Synergy journey and collaborations with new generations. Examples include the 2011 première of Carl Vine's *Three Pomegranates* for two marimbas, frequent performances of Nigel Westlake's marimba quartet *Kalabash* (2004),[4] and multiple commissions from Amanda Cole and Alex Pozniak, as shown in Table 3.2. In the most recent period of commissioning for the group, the number of non-notated works has increased, and in some cases scores or programs listing these works don't exist. These works appear in projects such as 1, 2, 3 (2010) and City Jungle (2011), both of which were designed as a through-composed performance of new repertoire and therefore had a single name encompassing a number of smaller works. Between 2009 and 2017, Timothy Constable replaced Askill as artistic director, and Cleworth reflects that:

> The approach and sound developed by the original members of Synergy continues even though none of the original members are in Synergy any more. This approach can be defined as open, collaborative and broad in scope.
>
> (Cleworth, 2013)

A series of major events, including the aforementioned 40under40, marked the 40th anniversary of Synergy in 2014. Titled Xenakis vs Pateras, the first of these events took place over two nights at Carriageworks in Sydney, and was performed by past and present Synergy members Timothy Constable, Bree van Reyk, Joshua Hill, Ian Cleworth, Mark Robinson and William Jackson. The first night featured Xenakis' *Pléïades* and the second the première of an hour-long commission for percussion sextet and spatialised electronics by Anthony Pateras; *Beauty Will Be Amnesiac or Will Not Be At All* (2013). This work represents the largest percussion ensemble work produced by an Australian composer to date in both

Table 3.2 Australian repertoire commissioned by Synergy Percussion since 2000

2000	David Hewitt	*Archipelago*	Drum kit soloist and sextet
2000	Ian Cleworth	*Fragments . . . Forms*	Percussion solo and quartet
2000	Synergy Percussion and Fritz Hauser	*Drumlines*	Various
2001	Michael Askill	*Spirals*	Quartet
2001	Michael Askill	*No Rest from the Dance*	Sextet
2001	Michael Askill	*Salome Ballet Score* *Salome's Entrance* *Salome's Dance* *Herod's Drunken Dance* *Salome and John the Baptist* *Soldier and Page*	Various (from *Salome* project)
c. 2001	Michael Askill	*Outer, Inner, Secret and Real* *Cloudswing* *Air and Other Invisible Forces Suite* *Ocean Dance* *Together Again* *Speed of Light*	
2002	Timothy Constable	*Waves 02*	Flexible
2003	Kelly Staines	*Touch*	Quartet
2003	David Hewitt	*Bjork Variations*	Septet
c. 2003	Philip South	*Snore*	Trio and tape
2003	Philip South	*Saru Kabuto*	Quartet
n.d.	Ben Walsh	*Chi*	Quartet
n.d.	Dale Barlow	*Roundabout*	Quartet (arrangement)
2004	Timothy Constable	*Waves 04*	Flexible
2004	Timothy Constable	*Etchings IV*	Solo
2004	Timothy Constable and Michael Askill	*Home Cosmography*	Quartet and electronics
2005	Synergy Percussion and Fritz Hauser	*Drumlines (2)*	Various
2005	Timothy Constable	*Waves 05*	Flexible
2006	Timothy Constable	*Waves 06*	Flexible
2007	Timothy Constable	*Waves 07*	Flexible
2007	Timothy Constable	*Suna*	Quartet
2008	Gerard Brophy	*The Book of Clouds*	Synergy, Taikoz and orchestra
2008	Timothy Constable	*Waves 08*	Flexible
2008	Synergy Percussion and Fritz Hauser	*Russell and Crowe*	Quartet (for *From Space and Time*)
2008	Bree van Reyk, Timothy Constable, Fritz Hauser and Michael Askill	*The Fives*	Any number of players (for *From Space and Time*)
2008	Synergy Percussion and Fritz Hauser	*The Dull Flame of Desire*	Quartet and electronics (for *From Space and Time*)

2009	Timothy Constable	*Waves 09*	Flexible
2010	Timothy Constable	*Percussion Symphony*	Percussion orchestra (for *Pulse:Heart:Beat*)
2010	Timothy Constable	*Ténéré*	Percussion ensemble (for *Pulse:Heart:Beat*)
2010	Timothy Constable	*Blue Candle*	Percussion ensemble (for *Pulse:Heart:Beat*)
2010	Timothy Constable	*Six Petals*	Percussion ensemble (for *Pulse:Heart:Beat*)
2010/13	Timothy Constable	*Cannons of Navaro*	Quartet (for *Pulse:Heart:Beat*)
2010	Bree van Reyk	*Duet with Blindfold*	Duo (for *1, 2, 3*)
2010	Bree van Reyk and Timothy Constable	*Cold Harbour*	Two percussion and electronics (for *1, 2, 3*)
2010	Bree van Reyk and Timothy Constable	*Music for Insects*	Duo (for *1, 2, 3*)
2010	Gerard Brophy	*Gethsemane*	Synergy Percussion and the Song Company
c. 2011	Gary France	*Soundscape: Morning*	Percussion ensemble
2009	Timothy Constable	*Last Waltz*	3 versions: quartet, trio and duo
2011	Carl Vine	*Three Pomegranates*	Marimba quartet
2011	Timothy Constable/ Synergy Percussion	*City Jungle*	Contains ten smaller works
2012	Alex Pozniak	*Groove Destruction*	Quartet (for *Emerging Composers*)
2012	Amanda Cole	*Intermetallic*	Quartet (for *Emerging Composers*)
2012	James Rushford	*Words Halfway to Intelligence*	Quartet with electronics (for *Emerging Composers*)
2012	Marcus Whale	*Puff*	Quartet (for *Emerging Composers*)
2012	Damien Barbeler	*Silver Lining*	Tabletop Trio
2013	James Humberstone	*Campanella*	Quartet (for *Bespoke for Air*)
2013	Julian Day	*Disco Demolition Night (What You Do to Me)*	Quartet (for *Bespoke for Air*)
2013	Kate Moore	*Sarabande*	Jaltarang Trio (for *Bespoke for Air*)
2013	Evan Mannell	*Calm Yourself, Restless Mind*	Quartet (for *Bespoke for Air*)
2013	Anthony Pateras	*Beauty Will Be Amnesiac or Will Not Be At All*	Sextet and electronics

instrumentation and duration. Pateras has a long history of collaborations with percussionists, with 11 prior percussion works, including *Transmutations* (2002), *Hypnogogics* (2005) and *Mutant Theatre* (2001–2008). Discussions between Pateras and Constable began following Synergy's 2011 presentation of *Pléiädes*, and the spatialised set-up of Xenakis' first sextet *Persephassa* became the point of departure for Xenakis vs. Pateras. Pateras states:

> This was one thing I borrowed from Xenakis . . . It's basically the stage plan from *Persephassa* (a circle around the audience), except there's a speaker between each player for the sound spatialisation, with the instruments from *Pléiädes*. So I guess it's a conglomeration of his two sextets on the level of basic materials.
>
> (Aronowicz, 2014)

Using a spatialised set-up in Carriageworks' open performance space allowed audience members to walk freely around the performers on both nights. On the first evening, it also allowed six additional performers (Rebecca Lagos, Eugene Ughetti, Yvonne Lam, Leah Scholes, Louise Devenish and Claire Edwardes) to join the ensemble at the conclusion of *Pléiädes* for an arrangement of *Mélanges* for 12 percussionists by Constable. Developing new interpretations of staple works in Synergy's repertoire have become a signature way to mark significant dates in the ensemble's history. For example, in 2004 the ensemble recorded two different interpretations of Westlake's *Omphalo Centric Lecture* to mark the group's 30th anniversary: a new realisation of the original and a longer version arranged by Philip South (Synergy Percussion, 2004).

In reflecting on Synergy Percussion's history, Constable suggests that there have been four or five distinct periods of work (Constable, 2014). The hallmarks of each of these periods were strongly connected with the interests of the performers active during each period, with an ongoing emphasis on collaboration and experimentation. Although Synergy's output spans a wide variety of musical styles, the thread that runs through all of them is the flexible approach to making new work, whether that is engaging with the practice of a visiting artist such as Hossam Ramzy or Trilok Gurtu, the aesthetic of Australian composers such as Westlake or Pateras, the interpretation of twentieth century masterpieces by Xenakis or Reich, or the sound and connection with the instruments themselves. Synergy describes their aesthetic as 'a world of sound with percussion at its heart' (Constable, 2017) and Constable cites the performers' approach as a key focus, an aesthetic passed on from former ensemble members. He remembers that 'Michael has this beautiful way of approaching an instrument as though it was the first time, this way of listening without taking anything for granted' (Constable, 2014).

As previously outlined, this approach has been cited as highly influential on the development of Australian percussion practices in part due to the significant number of percussionists that have been involved with Synergy as core members, guest artists or associate artists. Founded in 1992, the Associate Artist program served to create professional development opportunities for young percussionists, and a number of associates including Alison Pratt, Bree van Reyk and Joshua Hill later became core members of the ensemble. Numerous guest percussionists from around Australia including John Clark, Iain Scotland, Peter Jacob, Claire Edwardes and Simon Barker (NSW), Eugene Ughetti, Matthias Schack-Arnott, John Arcaro and Leah Scholes (Victoria), and Louise Devenish and Paul Tanner (WA) highlight further reach and connection to the Australian percussion community.

Synergy also provided a creative and organisational springboard for two other ensembles to develop adjacent to Synergy's core activities: Japanese-Australian drumming ensemble Taikoz and experimental percussion/electronic songwriting trio Moth. Although each group is artistically distinct, all three ensembles are administrated by the umbrella organisation of Synergy Percussion Limited, and have co-produced work together such as *pulse:heart:beat* (2013).

Taikoz

The 1991 *Matsuri* project was the first collaboration between Ian Cleworth and Riley Lee. Ten years earlier in 1982, following completion of his study in Adelaide, Cleworth was awarded an Australian Japan Foundation scholarship to study taiko in Yamanashi Prefecture, Japan. Cleworth had relatives living in Tokyo, who had introduced taiko to him by sending LP recordings of taiko performer Sen Amano. Upon receipt of the scholarship, Cleworth's uncle facilitated a formal introduction that enabled Cleworth to join Amano's Japanese students in daily *wadaiko* lessons and ensemble sessions for three months. He reflects that at the time he 'had no idea what I was getting myself in for' (Cleworth, 2013), and attributes his interest in exploring and understanding non-western percussive traditions in part to Katherine Ellis, who lectured compulsory undergraduate ethnomusicology courses at Adelaide University in the 1980s. Cleworth returned to Australia with a profound appreciation for *wadaiko* and basic taiko skills. Initially, Cleworth did not plan to bring these skills into his percussion practice in Australia, believing that few links between western percussion and taiko performance practice could be made (Cleworth, 2013). When Cleworth met Riley Lee in Sydney, this began to change. As the first non-Japanese person to achieve Grand Master status of the shakuhachi and the first non-Japanese professional taiko player in the world with the renowned ensemble Ondekoza in the 1970s, Lee's history with taiko presented another perspective. The music of Ondekoza became a major influence on Taikoz as Lee shared ideas about taiko technique, approach and philosophy he developed during his four years with Ondekoza on Sado Island in Japan.

Following *Matsuri*, Cleworth's ongoing interest in *wadaiko* led to the inclusion of a number of short taiko pieces in Synergy's programming in the mid-1990s, arranged by Amano for the group. Cleworth reflects on the context of taiko within Synergy at the time:

> It was a natural extension for a lot of percussionists then to literally introduce the sounds and rhythms and pieces from other cultures. But you start

> to realize as you get deeper into things that are starting to mess with other cultures' music that while it might have good intentions . . . it does have a whiff of cultural imperialism. We had permission to perform that music, so it had some level of integrity there. But I think it was within the general climate of a lot of percussion groups and percussion players introducing music and instruments from other cultures to their work.
>
> (Cleworth, 2013)

As his interest in exploring *wadaiko* and creating new work grew, Cleworth recognised that in order to develop an original artistic practice that challenged the performers and the art form with integrity that 'a group made up specialist players devoted to the music, practices, styles and techniques of *wadaiko* would be required' (Cleworth, 2017).

In 1997 Cleworth and Lee founded Taikoz, an ensemble that occupies a unique place in Australian contemporary music. After 20 years of study, collaboration and creation, 'Taikoz are respected as innovators and perpetuators of taiko tradition internationally' (Clark, 2014), and in 2016 Cleworth and Lee each received a commendation from the Japanese Foreign Minister for contribution towards promoting friendship and goodwill between Australia and Japan. Taikoz are sometimes misunderstood to be performers of traditional Japanese music, in part due to problematic media descriptions as Australia's leading Japanese drumming ensemble (Fulton, 2011). Although Japanese drumming is undoubtedly a major influence, Taikoz is better understood as a group that primarily develops new music by Australian and Japanese artists informed by the traditional and modern techniques required to perform Japanese instruments such as odaiko, shimedaiko and shakuhachi. Cleworth reflects on a performance of Maki Ishii's contemporary taiko ensemble work *Monochrome* (1976) as a pivotal moment in the birth of Taikoz:

> I felt a real connection because they weren't playing traditional music as such, but a new piece of percussion music . . . on the taiko but with all the stylistic approaches associated with taiko performance. So that was very exciting. I thought this could be a meeting point where percussion meets taiko.
>
> (Cleworth, 2013)

The place of Taikoz at the nexus of contemporary percussion and taiko practices is further articulated thus:

> While we use Japanese instruments as our mode of expression, we are Australian musicians working in Australia and draw on a lot of other aspects of our culture here. We're respectful of the roots and traditions of this music but we have to be true to who we are.
>
> (Cleworth, 2013 quoted in Clark, 2014)

This desire commenced with a programming model not dissimilar to Synergy's, where well-known pieces of existing repertoire were programmed alongside new works by Australian composers, often composers from within the ensemble. Their first performance at the Sydney Town Hall reflects this, including some traditional taiko music, Ishii's *Monochrome* and several works by Taikoz performers. Although their works were largely well received, soon the ensemble sought to make new work with greater depth and stronger technique. In 2001, they began what were to become regular study trips to Japan as an ensemble.

Cleworth describes Taikoz's two principal areas of creative endeavour as a framework upon which to build an original form of taiko (Cleworth, 2017). The first, 'the creation of new taiko music by Australian and international composers, is based upon the second: the study and (sometimes) performance of drum and flute music from Japan'. He continues: 'Taikoz's aim is not simply to transplant one culture's music into another's, but to allow a new form of music to grow upon a sure foundation of understanding the aesthetic principles and applying the techniques of *wadaiko*'. Like other members of Taikoz, Cleworth has developed an approach to this in his music for the ensemble, shown in Table 3.3.

Pursuing these goals has led to collaboration with Japanese *wadaiko* and western art musicians. Collaboration has been a centrepiece of Taikoz's practice, including projects with influential Japanese taiko artists Kodo, Eitetsu Hayashi, Kenny Endo and Kaoru Watanabe, as well as the Bell Shakespeare Company and choreographer Meryl Tankard in Australia. In 2011 the ensemble began developing their first full-length original production together with Sydney-based choreographer Anandavalli Sivanathan and her South Indian classical dance company Lingalayam. Titled *Chi Udaka*, this work was premiered at the 2014 Sydney Festival before touring in Australia and India. Taikoz has also performed taiko concerti with the Sydney, Melbourne, West Australian and Queensland Symphony Orchestras and the Dresden Sinfoniker (Germany).

Although the influence of Japanese taiko artists and contemporary Japanese approaches to *wadaiko* music making have been significant in the development of Taikoz, local influences are equally a part of the group's identity. The training of most of the current ensemble members in contemporary percussion in Australia has given the ensemble some unique characteristics. For example, Taikoz is the only taiko group in the world with all members able to read and use western musical notation. Lee notes that while this in itself is 'not a requirement of, or a guarantee for a vibrant and successful musical group, it does allow Taikoz to learn quickly and perform well, pieces that are as diverse as they are numerous' (Lee in Cleworth, 2017). Taikoz has created over 30 original *wadaiko* works, including several that incorporate diverse instruments such as shakuhachi, koto, marimba, saxophone and didjeridu. The combination of traditional physical discipline, Japanese musical influences and contemporary percussion approaches to composition evident in Taikoz's methods has enabled the ensemble to developed their own unique hybrid art form that could be described as an Australian contemporary approach to taiko.

Taikoz has built a community around this Australian contemporary approach to taiko, evidenced by their growing repertoire, strong touring program and a large taiko school. Lee reflects on his earliest performances with Cleworth in the 1990s,

Table 3.3 Works composed by Ian Cleworth for Taikoz

1996/2001	Ian Cleworth	*Yuuraku (Night Music)*	Solo shakuhachi, Odaiko and Rin
1997	Ian Cleworth	*Kenjo*	Taiko septet
2000	Ian Cleworth	*Dedication*	Shimedaiko Quintet
2001	Ian Cleworth	*Ota-i-ko*	Large taiko ensemble (7+ players)
2003	Ian Cleworth, Riley Lee and Matthew Doyle	*Shifting Sands*	Taiko sextet, solo shakuhachi, didgeridoo and percussion
2004	Ian Cleworth	*Shinju (Pearl)*	Odaiko Quintet
2004	Ian Cleworth	*Towards the Crimson Sky*	For Bass Koto and percussion
2007	Ian Cleworth	*. . . the peals of thunder*	Taiko septet
2007	Ian Cleworth, Riley Lee, Timothy Constable and Taikoz	*Kaidan*	Taiko ensemble (6 players), solo shakuhachi, electronics
2007	Ian Cleworth	*The Gathering*	Taiko ensemble (6 players), shakuhachi
2009	Ian Cleworth	*Clouds Wander*	Solo Soprano Saxophone and taiko trio
2009	Ian Cleworth	*House Music*	Taiko sextet
2011/2012	Ian Cleworth	*Towards the Crimson Sky*	For Bass Koto solo, taiko ensemble, shakuhachi, shinobue, percussion and audio playback
2012	Ian Cleworth	*Awakening*	Taiko ensemble (7 players)
2012	Ian Cleworth	*Of the Fields*	Katsugi-okedo quintet
2012/ 14	Ian Cleworth, Riley Lee, Aruna Parthiban and John Napier	*Chi Udaka*	Taiko quintet, solo shakuhachi, cello, voice and percussion
2013	Ian Cleworth	*The Silver Pavilion*	Bass Koto solo, Koto ensemble and solo percussion
2013	Ian Cleworth and Riley Lee	*Cascading Waterfall*	Solo shakuhachi, voice, taiko, percussion and audio playback
2014	Ian Cleworth	*Misra*	Taiko quintet, 2 shinobue

and the influence on the Australian context of Taikoz's roots. He reflects on their second performance, which was for the opening of an aluminium smelter in Gladstone, Queensland: 'Ian performed on his newly commissioned *ōdaiko*, which he had had made from a single Tasmanian eucalyptus tree. The beginnings of Taikoz could not have been more quintessentially Australian' (Lee in Cleworth, 2017).

Ensemble activities have been central to percussion activity in Sydney for decades through the work of Synergy, Taikoz and the creative environment at the Sydney Conservatorium of Music fostered by Daryl Pratt and Richard Miller. The diverse Sydney percussion scene experienced further growth post-2000 with a number of percussion duo projects. For example, Daryl Pratt and Alison Pratt established Match percussion duo in 2001, an ensemble that primarily performs marimba and vibraphone repertoire composed or arranged by Daryl Pratt, including *Water Settings* (2005) and *Ngarin* (2006). Recently, the duo has collaborated with hand percussionists Jess Ciampa and Gabriel Fischer to develop further new repertoire for percussion quartet. In addition to Taikoz projects, a number of the Taikoz members have maintained a contemporary percussion practice with which they contribute to the wider Sydney percussion scene. For example, between 2002 and 2013 Kerryn Joyce and Kevin Man were active as Karak Percussion, composing a number of marimba and vibraphone works that are frequently performed by percussionists around Australia, such as *Shichi-Karak* (2002), *Enso* (2003) and *Lullaby* (2003). Although Karak was originally Brisbane-based, the ensemble is included here as following relocation to Sydney, their output expanded to include a long-term shakuhachi and percussion collaboration with Riley Lee in Shak-karak.

Although percussion ensemble activity has been at the heart of Sydney's percussion identity, other vital activity adjacent to the collective ensemble environment in the 1980s and 1990s has played a crucial role in developing practices in this city. In particular, projects by percussionist-composer Graeme Leak and later, soloist Claire Edwardes made significant contributions to Australian percussion. Both were associated with Synergy Percussion in the 1980s and 1990s, respectively, albeit in different capacities, and both engaged in various other influential contemporary percussion activities. At times, this has intersected with the activities of Synergy Percussion or its members, therefore this output merits discussion here.

Graeme Leak

The work of percussionist and composer Graeme Leak (b. 1960) incorporates composition, community music making and instrument building. Leak's diverse practice has enabled him to create a large body of repertoire that occupies a unique place in Australian percussion history, encompassing solo percussion, community percussion ensemble, junk percussion, found objects and sound installation oeuvres. Although Leak has spent substantial time in Melbourne and Brisbane during his career, it commenced in Sydney and thus is discussed in this chapter. Leak's interest in contemporary percussion performance was ignited when he was a secondary student at the Sydney Conservatorium High School, which at the time was the highest level of training available for percussionists in New South Wales. When Leak heard Keith Humble's Melbourne-based Australian Contemporary Music Ensemble (ACME) on the radio he decided to pursue a career in contemporary music and wrote a letter to the ACME musicians:

> I heard The Australian Contemporary Music Ensemble on the radio when I was about seventeen and I was blown away by what I heard. I'd never heard anything like it.
>
> (Leak, 2012)

> Simone de Haan was on trombone and Christian Wojtowicz was on cello. I heard these guys on the radio and I wrote to them and sent them a photo and said 'I want to join your band'. Simone still remembers the letter arriving and they all said, 'Check this kid out from some school who wants to join the band' and they all had a bit of a chuckle.
>
> (Leak, 2013)

Leak reflects that although this letter didn't result in immediate work with the ACME, he began working as a percussionist in other ensembles prior to graduation from high school in 1978. After graduating from high school, Leak joined a handful of percussionists who were able to embark on full-time freelance careers. For many percussionists active during this period, the primary options for employment lay in education or orchestral performance. Leak's early performance work during this period included casual orchestral percussion work with the Australian Chamber Orchestra (1979–1988), Queensland Symphony Orchestra (1981–1982) and Sydney Symphony Orchestra (1978–1990). However, it was his work with contemporary chamber music ensembles the Seymour Group (1979–1986), the Australia Ensemble (1979–1986), Martin Wesley-Smith's electronic and computer collective WATT (1986–1988), Touchwood, (1983), Synergy Percussion (core member 1986–1987, guest 1980–1985) and Flederman (1979–1988) that expanded the diversity of his output at an early stage of his career. He remembers this as a time when there was a high volume of work available in Australia:

> Samplers hadn't been invented and there was lots and lots of work. I started working professionally when I was still at high school and just carried on. I never did any tertiary study myself, I went straight out as a percussionist after school and I was very busy in Sydney as a freelance player.
>
> (Leak, 2012)

Of these ensembles, repertoire performed by both Touchwood and Flederman in particular demanded virtuosic percussion performance and thus have been cited as influential in the growth of Australian percussion in the 1980s by a number of percussionists (Askill, 2013a; Cleworth, 2013; France, 2013; Lagos, 2013; Piper, 2013; Pye, 2012; White, 2012). In 1980, a few years after he wrote to de Haan as a school student, Leak was invited to join Flederman. Founded the previous year by Simone de Haan (trombone) and Carl Vine (composer) following the disbanding of ACME, Flederman adopted a flexible ensemble format, sometimes performing as the Flederman Trio and at other times as the Flederman Sextet by including Graham Hair on piano and synthesizer, Geoffrey Collins on flute and either Georg Pederson or Christian Wojtowicz on cello. When the larger ensemble required two percussionists, Leak was joined by Michael Askill. Leak cites his Flederman colleagues as integral to his pursuit of contemporary percussion. For example, regular collaboration between Askill and Leak through Touchwood, Synergy Percussion and Flederman resulted in an inevitable cross-pollination of influences and ideas. For example, when Askill gave the Australian première of

Child of Tree (1975), John Cage's solo for amplified plant materials in 1985, Leak was the 'cactus technician' (Leak, 2013). Inspired by the sounds produced, Leak later used an amplified cactus in his first major creative production *Homeboys* in 1988. Additionally, experimentation through collaborative improvisation was a key component of Flederman's relentless pursuit of new musical ideas and development of repertoire. The ensemble moved between cities during the 1980s, and Leak acknowledges the important role of improvisation during Flederman's time in Queensland when reflecting on the group's rehearsal methods:

> We used to improvise every day, sometimes for two or three hours. We did about 25 concerts in regional Queensland and at the Conservatorium [in 1981 and 1982] and we used to fly down to Sydney and put together the Flederman Sextet.
>
> (Leak, 2013)

As Synergy Percussion had shown in a percussion ensemble context, the music programmed by Flederman in particular demonstrated the capabilities of percussion to composers, percussionists, improvisers and the concert-going public in a chamber music context in a way that hadn't yet been seen in Australia. In Flederman, percussion was treated as equal to the other instruments in the ensemble and the result was an ensemble with demanding, virtuosic percussion parts at the heart of the ensemble's sound. Leak remembers the increased interest Australian composers had in the sound world available to them through the use of percussion at the time and reflects on the huge set-ups of instruments required: 'It was the '80s and composers were going crazy with percussion. I mean, the percussion setups needed truckloads of percussion for one player, it was very multi-instrument-one-player [sic] work' (Leak, 2012). An example of 'composers going crazy with percussion' can be found in the multiple percussion solo composed for Leak by Keith Humble. *Sonata for Percussion* (1986) also known as *Module #1 for Percussion*, called for marimba, vibraphone, glockenspiel, six crotales, brake drum, two tam-tams, six suspended cymbals, sizzle cymbal, six temple blocks, woodblock, log drum, six wood drums, cowbell, six water drums and six tubular drums. The tubular drums are unusual instruments that were built by Leak specifically for this work. Modelled after a set of tub drums made by Keith Humble and Jean-Charles François for the La Trobe University Improv Lab,

> the tub drums were PVC pipe drums with yoghurt tubs as caps. There was a size of [yoghurt] tub that fitted a size of PVC pipe perfectly – from memory the tub was cut in half and you slipped the bottom bit (using the base as the 'drumhead') onto the end of the PVC pipe. Sounded exactly the same as thong-o-phone, but could be played with stix [sic].
>
> (Leak, 2014)

The percussion-centred instrumentation of Flederman would later influence other chamber ensembles such as Nova Ensemble in Perth and Ensemble Offspring in

Sydney. Pye remembers Flederman as a particularly unique ensemble in contemporary chamber music at the time: 'They were unbelievable. They were a revelation when they started and the repertoire they created was unlike anything from anywhere else' (Pye, 2012).

Flederman placed a strong emphasis on interaction between composers and performers and regularly commissioned works from composers within the ensemble, such as Vine and Hair, and their Australian colleagues, including David Lumsdaine, Anne Boyd, Martin Wesley-Smith and Andrew Schultz. Every concert involved the première of new works and their high volume of new repertoire contributed to the level of influence Flederman had on their peers. The constant engagement with composers via Flederman was important for Leak as a young percussionist: '[Flederman] certainly was the most powerful influence on my early years because hanging around with those people also meant that I was connecting with the networks of composers and experimental improvisers and electronics and it was a very exciting time' (Leak, 2013).

In addition to his work with chamber ensembles, Leak is also credited with being the first Australian percussionist-composer to compose a multiple percussion solo. Leak's work. . . *And Now for the News* (1984) (see Example 3.1) is often thought to be the first Australian percussion solo, however, it was preceded by at least ten works (including Leak's own 1978 work *Just Testing*) three of which featured multiple percussion set-ups.[5] Like *Omphalo Centric Lecture*, . . . *And Now for the News* is an example of a work composed for a Touchwood performance

Example 3.1 An excerpt from . . . *And Now for the News* by Graeme Leak showing the speech rhythms that became the basis for the score.

that has found a lasting place in the Australian percussion canon. The intent behind this work is described in a 1987 program note:

> This piece was inspired by a broadcast of Vietnamese news on Sydney's radio station, 2EA. The elegant speech-rhythm seems to have an underlying pulse, so I recorded subsequent news broadcasts and overlaid them with a constant drum beat. I then transcribed the resulting rhythms and these transcriptions are in fact the score. Using this material as a basis for improvisation, the performer is free to play with or against the tape, to play solo or let the tape play solo, and anything else he or she may feel is appropriate! Although I have always played the piece on an array of drums, the instrumentation is variable.
>
> (Leak, 1987)

In addition to composing and commissioning percussion solos for performance in chamber ensemble programs, in 1987 Leak toured a solo percussion show consisting entirely of Australian repertoire to universities around Australia. This was the first time an Australian percussionist had embarked upon a series of solo performances and the repertoire included premières of works by Leak and arrangements of Touchwood repertoire for one percussionist. An emphasis was placed on works that combined electronic and acoustic percussion instruments, and at the time, this was unusual for solo percussion performances.

The complete program Leak presented during this tour included five instrument-and-tape pieces, namely Martin Wesley-Smith's *For Marimba and Tape* (1983), Graeme Leak's. . . *And Now for the News* (1984), Greg Schiemer's *Music for Shreelata* (1986), Carl Vine's *Percussion Concerto* (1987) and a solo version of Nigel Westlake's *Omphalo Centric Lecture* (1984) with pre-recorded marimbas by Askill. Also on the program was Phil Treloar's multiple percussion solo *Percussion Music One* (1985) and Greg Schiemer's *Monophonic Variations* (1986) for Octapad, Datum computer and Akai sampler. Leak collaborated with each composer on these works prior to the tour, as all except *Omphalo Centric Lecture* were composed for Leak. When Leak moved to Melbourne in 1991, an opportunity to develop a new show of original solo percussion material arose within the Astra Chamber Music Society concert series.[6] Titled *From Africa to the Kitchen Sink,* it was first performed at Elm Street Hall in North Melbourne on 28 July 1991, as shown in Figure 3.4. This project provided a platform for Leak to present ideas he had begun to explore in various collaborative and interdisciplinary projects. For example, Leak's quest for new sounds had led to the regular feature of junk percussion, toy instruments and found objects in his work, and this was exemplified in *From Africa to the Kitchen Sink*. Table 3.4 lists the complete program of *From Africa to the Kitchen Sink*, and lists everyday urban objects such as a briefcase, a sink, and a salad bowl amongst the instruments required. When Leak began touring *From Africa to the Kitchen Sink* to national and international festivals, such as the Percussive Arts Society International Convention in 1995, it was renamed *The Art of Noises* to reflect this instrumentation.

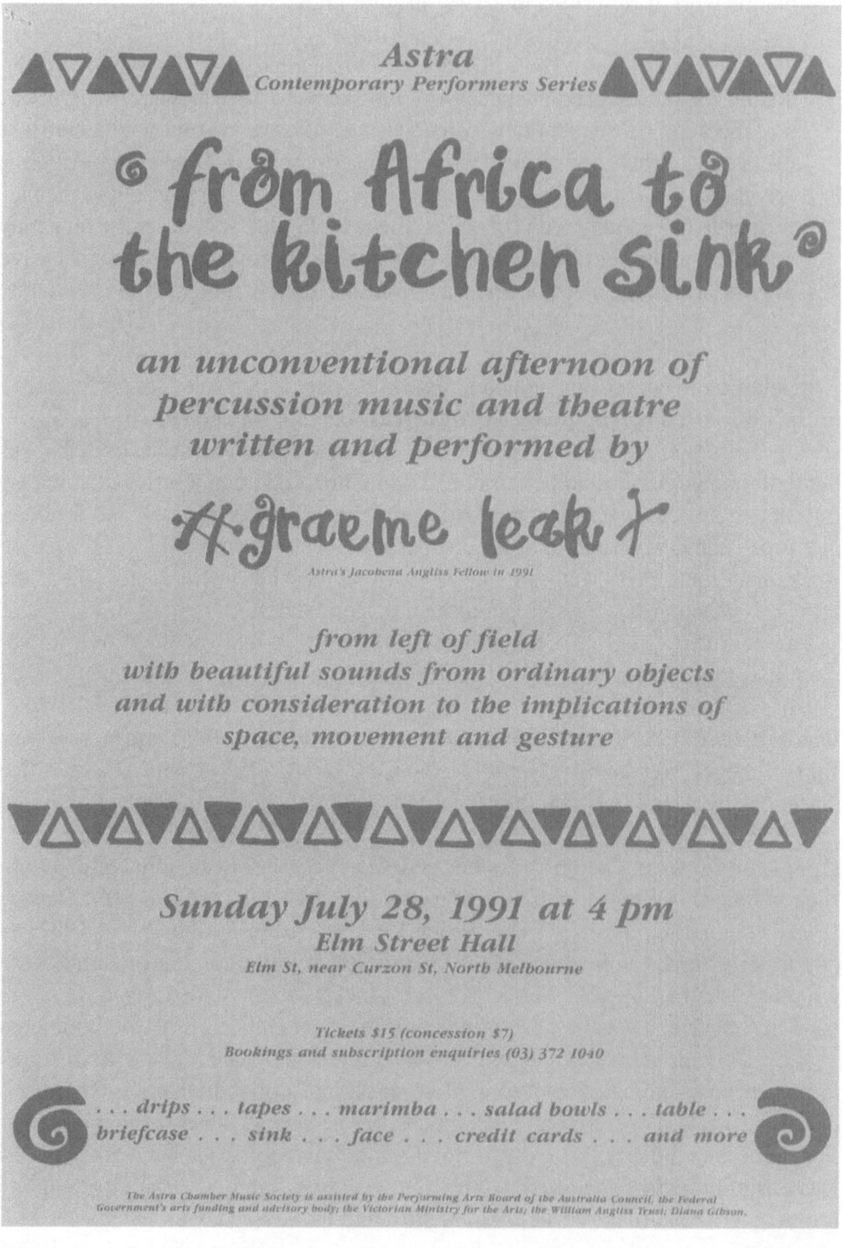

Figure 3.4 Poster advertising the 1991 première performance of *From Africa to the Kitchen Sink*.

Source: Artwork by Caroline Gyucha.

Table 3.4 From Africa to the Kitchen Sink program

Date	*Composer*	*Work*	*Instrumentation*
1991	Jon Hassell	*Flash of the Spirit*	Recorded music played prior to the show commencing
1991	Graeme Leak	*A Sad Song* (later renamed *Nobody Knows*)	Marimba and melodeon
1984	Graeme Leak	*. . . And Now for the News*	Drums, tape and Vietnamese voices taped from a radio 2EA broadcast, Sydney
1991	Graeme Leak	*I Love Jazz*	Inspired by African storytellers
1991	Graeme Leak	*In the Morning*	Marimba and tape
1989	Graeme Leak	*Spirit Dance*	Mbira, marimba, talking drum and tape
1989–1991	Graeme Leak	*The Briefcase*	Desk, briefcase, floor, face
1990	Graeme Leak	*The Sink*	Sink, tape
1991	Graeme Leak	*The Water*	Sink, salad bowl, tank
1991	Graeme Leak	*Yo Yo Man*	Marimba, feet and voice

Leak continued to draw inspiration from some of the concepts explored in *From Africa to the Kitchen Sink*, such as *The Briefcase*, and everyday urban objects featured in Leak's solo, ensemble and educational projects into the 1990s. Figure 3.5 shows Leak performing his semi-improvised work *Drumming Bunnies*. In this work, the soloist improvises a duet with a roboticised toy Santa Claus, an Energiser Bunny and a toy koala that plays miniature bass drum and cymbals. Each toy is fitted with midi triggers, and the speed of each toy's instrument, and therefore the tempo of the duet, is determined by an off-stage operator.

The use of everyday objects extended to the design and construction of instruments in the 1980s. For example. The Salad-Bowl Drums were Leak's urban Australian response to African water drums, comprising different sized wooden salad bowls floating in a large Perspex bowl filled with water. Light wooden salad bowls were easily purchased in supermarkets and homewares stores at this time, and today can be found primarily in second-hand shops. The Salad-Bowl Drums preceded the Bowl Chimes in 1984, which were made by attaching a metal chime to a bowl. Leak's bowl instruments were followed by the String Can in 1992 (Leak, 2003: 138). Constructed from a broom handle, a tin can and wire tuned by guitar fittings, the string can was used by Leak for solo pieces and in his ensemble, The Graeme Leak String Quartet, shown in Figure 3.6. The quartet included five players: composer Tim Dargaville, Tania Bosak, Phil McLeod, Stuart Campbell and Leak. Leak describes the intent behind the String Can thus:

> I'm a sound junkie and I'm always on the lookout for interesting sounds. I listen to a lot of traditional musics from different cultures around the world and I quite often hear beautiful instruments that we don't have in our Western

Figure 3.5 Graeme Leak performing *Drumming Bunnies*.

Source: Photograph supplied by Graeme Leak.

Figure 3.6 The Graeme Leak String Quartet, 1996: Graeme Leak, Jane Fisher, Tim Dargaville, Phillip McLeod.

Source: Photograph supplied by Graeme Leak.

> orchestras. Some of the Indian violins and the African violins are just so gorgeous and so much more interesting to me than the regular violin. So I set about to make them for myself. I just experiment and put things together and try and find that sound and the string can is a case in point. I was very interested in the sound of the berimbau but it looks like it comes from the bush and I didn't want to make something that looked like it came from the bush. I wanted to make something that looked like it came from Bunnings, because Bunnings is my 'bush'. I like to reference the suburban materials that we've got. I like junk and found, I like noise and I like bringing those crossover elements into an instrument. I love the way the Africans have rattles on their instruments and dirty up the sound. So, I guess I just follow the sound that I'm seeking. I hear a sound I want to make and I set around trying to put together an instrument to create it.
>
> (Leak, 2012)[7]

Between 1991 and 1999, Leak was also engaged as a lecturer in performance at the now defunct Music Department at La Trobe University. The use of everyday urban objects for improvisation was a significant part of Leak's teaching practice at La Trobe, particularly in the Improv Lab established for exploration, improvisation and composition by Keith Humble and Jeff Pressing in the 1970s when the La Trobe Music Department was founded. In addition to growing the existing instrument collection in the Improv Lab by adding everyday objects, Leak designed a project-oriented performance course called the D Stream that Leak's book *Performance Making* (Leak, 2003) is based upon. In 1998, Leak took up the position of Head of Music, however the Music Department closed in 1999. Although Leak had taught at a number of Australian universities, including the Sydney Conservatorium, he chose to pursue other professional avenues following the close of the La Trobe University Music Department. During his time teaching at the Sydney Conservatorium, he assisted in establishing percussion ensemble B'Tutta for Musica Viva in Schools in 1990. B'Tutta's original membership included Leak's students Graham Hilgendorf, Fatima de Asis, Cameron Gregory and David Hewitt and the ensemble remains active with the Musica Viva In Schools program nearly 30 years later, albeit with a changed membership.

Leak continues to be an active creator of unique percussion projects that incorporate unusual instruments and repurposed everyday urban objects. This includes composing and performing internationally with The Spaghetti Western Orchestra, whose theatrical shows featuring the music of Ennio Morricone presented with the aid of comedy and sound effects produced by objects such as coat-hangers, apples, rubber gloves, nail clippers and cornflakes, grew from small performances in Melbourne cafes to major international performances, including the BBC Proms in 2011. His major works cover a broad spectrum of musical disciplines including interdisciplinary projects such as *Ringing the Changes* (2010), a collaboration with physical theatre company Strange Fruit for aerial bell ringers and percussion duo, and the direction of his large-scale community projects such as *The Musical Fence* (Queensland, 2003), *The Incredible Hulls* for orchestra of boats, drummers and brass band (Tasmania, 2005) and *Drumming the Light* for percussion ensemble of any size (Western Australia, 2011). These projects are remarkable for their level of community engagement, often evidenced in the number of performers involved. For example, the première of *Drumming the Light* included an ensemble of 243 players, inclusive of professional, student and community percussionists.

Solo percussion in the 1990s

Towards the end of the twentieth century, percussion solo activity began to grow in the wider Australian music scene, aided by the success of percussionists in the widely publicised ABC Symphony Australia's Young Performer Awards. In 1981, South Australian John White became the first percussionist to win this award, then known as the ABC Instrumental and Vocal Competition (Atherton, 1990: 106).[8] He was followed by Alison Pratt (nee Eddington) in 1995. Pratt was a Synergy Percussion Young Artist in 1994, and following her win of the Young Performers Awards in 1995 she went on to spend 15 years as a core member of the ensemble. Five years later, Claire Edwardes (b. 1975) also won the competition, now known as the ABC Symphony Australia Young Performer Awards, and was the first to continue on the trajectory of solo performance following the competition. Shortly after however, Edwardes left Australia to study in the Netherlands and subsequently won a second international solo competition; the Tromp Percussion Competition in 2000. Prior to her departure, Edwardes was a Synergy Percussion Associate Artist, and a founding member of contemporary chamber ensemble Ensemble Offpsring. After spending time studying and working in Amsterdam, Edwardes returned to Australia in 2006. As the majority of Edwardes' work in Australia after this time falls beyond the dates of this study, only a brief outline of her work is included here. Edwardes reflects that since her return to Australia in 2006 and her career has progressed, she has found herself able to contribute to Australian percussion as an individual primarily through her artistic directorship and performance with Ensemble Offspring. This ensemble has premiered over 200 new works, many of which feature percussion, and consciously champions Australian repertoire. Engaging with repertoire by her composer colleagues in this way led to the 'gradual realisation that I have a role to play, in a way. I can help develop the repertoire with the composers, making new pieces with them' (Edwardes, 2013). Like Seal, Askill and Leak (and others to be discussed in subsequent chapters), Edwardes has been active in commissioning solo percussion repertoire from Australian composers including Alex Pozniak, Damien Ricketson, Gerard Brophy, Stuart Greenbaum, Amanda Cole and Matthew Shlomowitz. Edwardes regularly performs these works and has documented a number of them on CD recordings (see *Coil*, 2007, *Flash*, 2011 and *One*, 2013b). Another area in which Edwardes' output is significant is in the ongoing commissioning of percussion concerti from composers since her return, including Matthew Hindson, Iain Grandage and Elena Kats-Chernin. It is pertinent to note, however, that Edwardes' commissions are not the only percussion concerti by Australian composers. A handful of others composed since 1980 also exist such as Carl Vine's

Percussion Concerto (1987), and more recently new concerti such as *When the Clock Strikes Me* (2006) by Westlake for Rebecca Lagos and Matthew Shlomowitz's *Listening Styles* for Eugene Ughetti (2014).

Although not entirely exhaustive, this chapter extensively documents a range of contemporary percussion output in Sydney between 1974 and 2010, with a particular focus on the activities of Synergy Percussion. This ensemble was highly influential in the development of the Australian percussion repertoire, and a number of works composed for this group in the 1980s and 1990s have entered both the Australian and international canon. Demonstrated here is the centrality of the percussion ensemble to the development of contemporary percussion repertoire and performance practices in Australia during this period. Ensembles and their performers served as a hub for collaboration with composers, and the resulting works have significantly influenced Australian percussion identity (to be discussed further in Chapter Seven) as perceived by local practitioners and the wider community. The influence of Synergy Percussion can also be seen in the output of ensembles active in other parts of the country in the 1980s and 1990s, such as Adelaide Percussions and Nova Ensemble, who feature in Chapters Four and Five, respectively.

Notes

1 Askill's recording of *Omphalo Centric Lecture* is available on 'Onomatopoeia' (Westlake, 1994).
2 The original version of *White Knight and Beaver* was commissioned in 1984 by American trombone and violin/viola duo Miles Anderson and Erica Sharp. The xylophone and marimba arrangement was the first of many different permutations of the work. Neither Wesley-Smith nor Synergy hold a record of the year the arrangement was made, but estimate it was 1985.
3 Communication with the composer has not been able to confirm further details of this work. However, given the period it was composed, it is likely that the work was a quartet.
4 Note that *Kalabash* is often cited as a Synergy commission because they have recorded and regularly perform the work, however it was commissioned by Australian National University student ensemble DRUMatix.
5 Peter Clinch: *Fugue for Percussion* (1970), Felix Werder: *Percussion Play* (1972), Carl Vine: *Façade* (1973), Moya Henderson: *Sekko For It's Water* (1976), Graeme Leak: *Just Testing* (1978), Paul Sarcich: *Two Side Drum Solos* (1979), Robert Lloyd: *Marimba* (1979), Mark Pollard: *In Context* (1980), Ross Edwards: *Marimba Dances* (1982), Martin Wesley-Smith: *For Marimba and Tape* (1983).
6 Astra was formed in Melbourne in 1951 as an orchestra of women musicians under the direction of Astra Flack, a violinist and conductor who had migrated to Australia from Lithuania. In 1958, George Logie-Smith became musical director, extending the orchestra to include male members and a larger wind section and founding the Astra Choir. In 1978, current director John McCaughey established the contemporary concert series of which *From Africa to the Kitchen Sink* was part.
7 Bunnings is an Australian hardware chain store similar to Home Depot (USA) or Homebase (UK).
8 Following this, White began a career as a medical scientist researching the muscular-skeletal problems of musicians, and building mallet percussion instruments with colleague James Bailey (Atherton, 1990: 106–107).

References

Adamson, J. 1994. "Synergy's One-Night Stand has Reached Its 20th Anniversary." *Sydney Morning Herald*, 28 July.

Appleby, R. 2012. *Women of Note: The Rise of Australian Women Composers*. Fremantle: Fremantle Press.

Aronowicz, A. 2014. "Anthony Pateras Alone." *Resonate Magazine*, 17 April. www.australianmusiccentre.com.au/article/anthony-pateras-alone

Askill, M. 1986. *Australian Percussion*. Great Island Records. Compact disc.

Askill, M. 2013b. *A Life in Music*. Personal document. Brisbane, Queensland.

Askill, M. 2017. "Composing with Ancient Sound Technology in the Twenty-First Century." In *Contemporary Music Review*, 36 (1–2): 96–101.

Atherton, M. 1990. *Australian Made, Australian Played: Handcrafted Musical Instruments from Didjeridu to Synthesiser*. Kensington: New South Wales University Press.

Beck, D. 2011. *Set in Stone: The Cell Block Theatre*. Sydney: University of New South Wales Press.

Beck, J. H. 1995. "Percussion Ensembles." In *Encylopedia of* Percussion, edited by J. H. Beck, 269–273. New York: Garland.

Brown, P. 1974. "Memorable 'Rug Concert'." *Sydney Morning Herald*, 12 August.

Clark, F. 2014. "TaikOz: More than Muscles in the Media." In *Into the Mix People, Places, Processes: Proceedings of the 2014 IASPM-ANZ Conference*, edited by J. Cattermole, H. Johnson, O. Wilson, 17–26. Dunedin: International Association for the Study of Popular Music.

Cleworth, I. 2017. "Journeys and Transformations for the Wadaiko in the Australian Cultural Context, Focussing on the Work of TaikOz." Paper presented at the Transplanted Roots Percussion Research Symposium, Brisbane, 23 September.

Constable, T. 2017. "Synergy Percussion." Accessed December 7 2017. www.synergypercussion.com

Covell R. 1967. *Australia's Music: Themes of a New Society*. Melbourne: Sun Books.

Covell, R. 1989. "Masterfulness and Doodle: The Healthy Irreconcilables of Australian Music." In *Anthology of Australian Music on Disc: Handbook*. Canberra: Canberra School of Music.

Dalton, G. B. 2006. "Nigel Westlake's 'Omphalo Centric Lecture': A Guide for Performance Including a Biography of the Composer and an Examination of the Different Versions of the Work." DMA diss., Columbus: Ohio State University.

DeHart, J. 2010. "Tap Routes: the changing role of the contemporary artist-percussionist." DMA diss., University of California San Diego.

Edwardes, C. 2007. *Coil: Australian Music for Solo Percussion*. Tall Poppies TP193. Compact disc.

Edwardes, C. 2011. *Flash*. Tall Poppies TP215. Compact disc.

Edwardes, C. 2013b. *One*. Tall Poppies TP223. Compact disc.

Fulton, A. 2011. "Australian Artists Let in on Ancient Dance." *Sydney Morning Herald*, 25 April. www.smh.com.au/entertainment/dance/australian-artists-let-in-on-ancient-demon-dance-20110424-1dt2s.html

Globokar, V. 1992. "Anti-Badabum." *Percussive Notes*, 31 (1): 77–82.

Harley, J. 2004. *Xenakis: His Life in Music*. New York: Routledge.

Ice, S. P. 2012. "The Percussion Quartet: A Chronological Listing and Performance Guide of Six Selected Works." DMA diss., Norman: University of Oklahoma.

Kanach, S., ed. 2010. *Performing Xenakis*. New York: Pendragon Press.

Kerry, G. 2009. *New Classical Music: Composing Australia*. Sydney: New South Wales University Press.

Knowles, J. 2009. "Setting the Scene: Developments in Australian Experimental Music since the Mid-1990s." In *Experimental Music: Audio Explorations in Australia*, edited by G. Priest, 9–35. Sydney: New South Wales University Press.

Leak, G. 1987. *Programme and Composer Notes: Graeme Leak – Percussion*. Perth: WA Conservatorium of Music.

Leak, G. 2003. *Performance Making*. Sydney: Currency Press.

Leak, G. 2012. "Sketching Impressions: Radio Interview with Eugene Ughetti." *3MBS Radio*, 18 September.

Ludgate, T., I. Cleworth, M. Askill, and C. Piper. 1989. *Synergy to Paris for French Bicentennial Celebrations*. Sydney: Musica Viva Australia.

Martin, P. J. 1996. "Synergy Percussion: Australian Percussion Music in Context." Master of Music thesis, Armidale: University of New England.

Murdoch, J. 1975. *Australia's Contemporary Composers*. Melbourne: Sun Books.

Paget, J. 2013. "Has Sculthorpe Misappropriated Indigenous Melodies?" *Musicology Australia*, 35 (1): 86–111. doi:10.1080/08145857.2013.761101

Piper, C. 2012. "The Percussion Section of the Sydney Symphony Orchestra." *DRUMscene*, no. 70: 88–92.

Schaefer, J. 2007. "More About Michael Askill." www.michaelaskill.com/cms/introduction/34

Schick, S. 1995. "Multiple Percussion." In *Encyclopedia of Percussion*, edited by J. H. Beck, 257–263. New York: Garland Publishing Inc.

Schick, S. 2006. *The Percussionist's Art: Same Bed, Different Dreams*, 1st ed. New York: University of Rochester Press.

Skinner, G. 1991. "CD Review: Synergy Percussion." *Sounds Australian*, 31: 46.

Sydney Symphony Orchestra. 2017. "Richard Miller: Principal." Accessed 8 December 2017. www.sydneysymphony.com/about-us/meet-the-musicians/percussion-and-timpani/richard-miller

Synergy Percussion. 2002. "Steve Reich: Drumming." https://itunes.apple.com/au/album/steve-reich-drumming/id552551262

Synergy Percussion. 2004. *Omphalo Centric Lecture*. Self-published. Compact disc.

Westlake, N. 1994. *Onomatopoeia*. TP047. Glebe: Tall Poppies. Compact disc.

Westlake, N. 2005. "The Percussion Music of Nigel Westlake: An Interview with Gary France." www.rimshot.com.au/article/the-percussion-music-of-nigel-westlake/

Personal communication

Askill, M. 2013a. Skype Interview with Louise Devenish, 7 February. Mullimbimby, New South Wales.

Cleworth, I. 2013. Skype Interview with Louise Devenish, 25 March. Sydney, New South Wales.

Constable, T. 2014. Live Interview with Louise Devenish, 4 June. Perth, Western Australia.

Edwardes, C. 2013a. Live Interview with Louise Devenish, 23 April. Sydney, New South Wales.

France, G. P. 2013. Skype Interview with Louise Devenish, 25 March. Canberra, Australian Capital Territory.

Lagos, R. 2013. Live Interview with Louise Devenish, 25 April. Sydney, New South Wales.
Leak, G. 2013. Skype Interview with Louise Devenish, 16 July. Melbourne.
Leak, G. 2014. Email Communication with Louise Devenish, 24 August.
Piper, C. 2013. Live Interview with Louise Devenish, 26 April. Sydney, New South Wales.
Pratt, D., and R. Miller. 2016. Skye Interview with Louise Devenish, 27 October. Sydney, New South Wales.
Pye, D. 2012. Live Interview with Louise Devenish, 24 September. Fremantle, Western Australia.
White, T. 2012. Live Interview with Louise Devenish, 9 October. Perth, Western Australia.

4 Contemporary percussion in Adelaide

Although Adelaide was the centre of significant precursory activity that laid the foundations for Australian contemporary percussion in the 1960s, it was not until 1978 that contemporary percussion truly commenced in Adelaide. Distanced from other hubs of cultural activity, percussionists in Adelaide developed unique approaches to contemporary percussion projects, and the 1980s were a prolific time for percussion in this city. Coupled with the rising profiles of existing ensembles discussed in previous chapters, the presence of contemporary percussion in the Australian musical landscape increased significantly around the country in the 1980s as emerging professional percussion ensembles established their identities in Adelaide. Chapter Four focuses on the South Australian repertoire commissioned by Adelaide Percussions, including a comprehensive list of works. This is followed by a discussion of experimental activity in the 1990s, inclusive of the output by Vanessa Tomlinson and Catherine Oates' through their duo Rhythm Method.

Ryszard Pusz and Adelaide Percussions (1978–2000)

As Richard Smith's students became active professional musicians in the 1970s, various facets of percussive music making became firmly embedded in different areas of the Adelaide music scene. South Australia again led growth in percussion education, with the addition of a percussion faculty to the Music Branch of the Education Department of South Australia (est. 1962) in 1974 (hereafter Music Branch). The Music Branch employed full-time percussion teachers Ryszard Pusz, Dennis Johnson and Ray Horne, three of the many South Australian percussionists whose first introduction to percussion ensembles came via Richard Smith. Together with Peter Cobb, who held a similar position at the Flinders Street Technical College School of Music (hereafter Flinders Street), they formed Adelaide Percussions in 1978.

Adelaide Percussions was based at Flinders Street, and the support provided by Flinders Street was crucial to the viability of the ensemble. Access to an instrument collection and rehearsal space was key to the group's survival in the beginning, but Adelaide Percussions' affiliation with Flinders Street via Cobb, and the connection with the Music Branch via the other ensemble members, provided another advantage: audiences. As all four members of the group worked within the field of percussion education, they were personally connected to a large specialist audience for their concerts. It was common for Adelaide Percussions to play to full houses whether they were performing in a small recital venue at the Flinders Street School of Music, or in the 1100-seat Adelaide Town Hall (Johnson, 2013; Timcke, 2013).

By the late 1970s, awareness of contemporary percussion was growing in Australia and inspiration was available to the Adelaide Percussions from a variety of local and international sources. Pusz, the ensemble's director, cites Smith's It's Percussion ensemble as a major inspiration for both Adelaide Percussions and the various other student and professional percussion ensembles he directed (Pusz, 2013). Like many other Australian percussionists, the members of Adelaide Percussions were inspired by Les Percussions de Strasbourg's 1971 and 1976 Australian tours, and their influence is apparent in the choice of name for the group. Additionally, in 1976, two years prior to Adelaide Percussions' formation, Peter Cobb received a Churchill Fellowship to study with Les Percussions de Strasbourg in France. Other visitors to Australia also had some level of influence, including George Gaber, who had visited Adelaide as part of his Australian tour in 1975. During his visit, Gaber held workshops and rehearsals of percussion ensemble repertoire, thereby introducing a great deal of new American repertoire to the percussionists who would later form Adelaide Percussions. In 1976 – the same year Cobb was in Strasbourg – Pusz undertook a few months of study in Indiana with Gaber. Aided by a grant from the Music Branch for the purchase of new percussion music, Pusz returned to Australia with a large collection of repertoire purchased in Indiana that Adelaide Percussions later used in their performances.

That the Adelaide Percussions format was a quartet rather than a sextet can be attributed to a number of factors including logistics, availability of players and the fact that Pusz and Askill had previously tested the quartet model as students with Ian MacDonald and Michael Holland in the mid-1960s via an unnamed student ensemble. It has been suggested that Synergy Percussion (then named The Sydney Percussion) were an additional influence on Adelaide Percussions, as the majority of the existing Australian percussion repertoire available to the group at the time had been composed for The Sydney Percussion's quartet arrangement. Former Adelaide Percussions member Ian Cleworth stated in a 1996 interview with Peter John Martin that: 'Ryszard started it [Adelaide Percussions] very consciously as Adelaide's answer to Synergy' (Martin, 1996: 35). Conversely, in an interview with this author Pusz stated that Adelaide Percussions was not formed with the intent to compete with Synergy, rather that the 'conscious response' was to celebrate the growing percussion activity in Adelaide via a professional percussion ensemble. Pusz suggests that although there was a perception amongst their peers that Adelaide Percussions was 'Adelaide's answer to Synergy' (Pusz, 2014), that was not the intent of the founding members. Due to the date of Adelaide Percussions' formation, the similarity of the ensemble's names and the friendly relationship between Askill and Pusz as students in Adelaide, it is reasonable to believe that an awareness of The Sydney Percussions made some level of influence on Adelaide Percussions inevitable.

Although The Sydney Percussion provided an existing source of Australian repertoire, Adelaide Percussions recognised the need to commission new works in order to establish their own identity. Successful pursuit of Australia Council funding in the 1980s enabled the commissioning of new Australian works for percussion ensemble, the majority of which were commissioned from South Australian composers. Adelaide Percussions performed up to six different programs annually that regularly featured the premières of their commissions. Like the Australian Percussion Ensemble before them, Adelaide Percussions commissions included percussion ensemble, multiple percussion soli and mixed chamber ensemble repertoire.

The 47 works for percussion ensemble and mixed chamber ensemble commissioned by Pusz and Adelaide Percussions since 1978 represents a significant contribution to the Australian percussion repertoire, and is shown in Table 4.1. As the third professional Australian ensemble to engage extensively in commissioning, this ascertains the trend of supporting local composers and the pursuit of their own identity in the form of a unique body of repertoire. Further, formation of close collaborative relationships between percussionists and composers was a key part of the creative process and this was occasionally referred to in the titles or structure of new works for Adelaide Percussions. Although composers always wrote for the familiar instruments in the group's collection at Flinders Street, extensive rehearsals were necessary in order to develop the material for performance, as many of the commissions for Adelaide Percussions contained techniques and concepts new to the members of the group. In addition to performing commissions, Adelaide Percussions gave South Australian premières to a number of existing Australian works for percussion ensemble, such as Graeme Leak's *Drum Song* (1989), and mixed chamber ensemble, such as Felix Werder's *A New Sphere of Activity* (1981) for percussion and piano and Douglas Kneehans' chamber ensemble work *Exiles* (1987).

Although a large body of repertoire was developed for Adelaide Percussions, the impact of this repertoire beyond Adelaide was limited as most works were only performed once by the group. Johnson reflects that their full-time education positions in combination with the frequency new compositions were presented to the group also contributed to works being only performed once (Johnson & Pusz, 2014). Although Adelaide Percussions enjoyed a strong audience, population necessarily restricted the frequency of performances in Adelaide. Further, awareness of Adelaide Percussions' repertoire was limited by the fact that many were not published at the time, and some works by Australian composers during this decade were not premièred in Australia, such as Keith Humble's *Etchings* (1988). A similar problem existed with another ensemble, Robert Lloyd and Musicians. Lloyd, a student of Pusz, wrote a number of percussion works for this ensemble, which was active between 1973 and 1995 (Lloyd, 2013). Although Lloyd composed no fewer than ten percussion ensemble works during this period, including *Bhakti* (1974), *Marimba* (1979), *Haiku* (1979), *Etude* (1979), *4 Drums* (1980), *Shift* (1982), *Sticks* (1985), *Principles of Simultaneous Rhythms* (1985), *Boo Bam Music* (1991) and *Earthed* (1992), the ensemble gave infrequent concerts and most of these works were performed once.

Table 4.1 Notated Australian repertoire commissioned by Adelaide Percussions

Year	*Composer*	*Work*	*Instrumentation*
1979	Robert Lloyd	*Etude on Kandian DrumRhythms*	Solo percussion
1979	Robert Lloyd	*Whirling Dance*	Percussion quartet
1980	Bozidar Kos	*Quartet*	Percussion, piano, viola and flute
1981	Eric Bryce	*Suite for Percussion Quintet*	Percussion quintet
1982	Atis Danckops	*Songs for Dark Hours*	Percussion, flute, guitar and mezzo soprano
1982	David Morgan	*Loss*	Percussion quartet
1983	David Morgan	*Voyage into Solitude*	Percussion and clarinets
1984	Atis Danckops	*A Different Kind of Jazz*	Percussion quartet
1984	Atis Danckops	*4 Studies for 4 Percussion Players*	Percussion quartet
1984	Felix Werder	*Fractured Fancies*	Percussion, viola and piano
1984	Sarah de Jong	*Without Limits*	Percussion, piano and clarinet
1984	Andre Oosterbaan	*All the Reed Will Say*	Clarinet and percussion
1984	Ralph Middenway	*Stone River*	Percussion quartet and solo baritone
1986	David Morgan	*Fun and Games*	Percussion quartet
1986	Larry Sitsky	*Diabolus in Musica*	Percussion quartet
c. 1986[1]	Ralph Middenway	*Mosaics for Reeds and Percussion*	Percussion and clarinet
1987	Tristram Cary	*Rivers*	Percussion quartet and two tape recorders
1987	Eric Bryce	*Jazz Suite*	Mallet percussion quartet
1987	Eric Bryce	*Richard's Rag*	Three xylophones, four marimbas, vibraphone
1987	Felix Werder	*Renunciation*	Viola and percussion with chamber orchestra (double concerto)
1987	David Harris	*Six Lyric Pieces*	Multiple percussion and piano
1987	Ralph Middenway	*Mosaics*	Percussion solo
1988	Keith Humble	*Etchings for Percussion Quartet*	Tuned percussion quartet
1988	Becky Llewellyn	*Early Spring*	Glockenspiel or vibraphone solo
1988	Peter Brideoake	*Dialogue for Two*	Percussion and clarinet
1988	Eric Bryce	*Concerto*	Marimba/vibraphone solo and Orchestra
1988	Douglas Knehans	*Dawn Panels*	Percussion quartet and clarinet
1989	Eric Bryce	*Six Pieces for Marimba or Vibraphone*	Marimba or vibraphone
1989	David Harris	*Chamber Concerto*	Marimba and percussion ensemble
c. 1989	Felix Werder	*Quinny on the Roof*	Percussion solo

Year	*Composer*	*Work*	*Instrumentation*
c. 1989	Peter Brideoake	*Percuszion*	Percussion and chamber orchestra
1990	Larry Sitsky	*The Phantom Drummer of Tedworth*	Percussion solo
1990	Raymond Chapman Smith	*Untitled 17.6.90*	Percussion quartet
c. 1990	Peter Brideoake	*Dialogue for 4*	Percussion, viola and clarinet
1991	David Morgan	*An Ittsy-Bitsy Collection for Marimba*	Solo marimba
1991	David Morgan	*Concerto*	Percussion solo and orchestra
1991	Tristram Cary	*Black White and Rose*	Solo percussion and tape (marimba, woodblocks, gongs)
1993	Neil Currie	*Sonata for Percussion Quintet*	Percussion quintet
1992	Becky Llewellyn	*Song Web*	Percussion solo
1993	Becky Llewellyn	*Sticks and Stones*	Percussion solo
c. 1998	Felix Werder	*Klänge und Gegenklänge*	Percussion and viola
2000	David Harris	*Concerto for Multiple Percussion and Percussion Orchestra*	Multiple percussion and percussion orchestra
c. 2003	Eric Bryce	*The Romantic Gypsy*	Percussion, violin and guitar
2004	Felix Werder	*Galgenlied*	Percussion and guitar
2004	Aleksander Pusz	*Primitive Complex*	Percussion quartet
2005	Ian Seaborn	*Wetlands*	Percussion duo
2007	Nigel Sabin	*Postcards from France*	Percussion duo

Where Adelaide Percussions differs from The Sydney Percussion and the Australian Percussion Ensemble is in their programming of international works. Adelaide Percussions chose not to play any of Les Percussions de Strasbourg's repertoire, instead complementing Australian commissions with North American works by percussionists Anthony Cirone and William Kraft, ragtime xylophonist George Hamilton Green, and marimbist Bill Molenhof. The featuring of primarily North American rather than European international repertoire further highlights the influences of Richard Smith and Ryszard Pusz's time in Indiana, and the ongoing influence of North American practices on the emerging Australian percussion scene. As previously discussed, Smith's pedagogical practice included xylophone ragtimes and American method books; thus, his students became familiar with this style of percussion composition, while also being exposed to It's Percussion arrangements and this in turn influenced the programming of Adelaide Percussions.

During the 1980s and 1990s, the core membership of Adelaide Percussions changed to frequently include tertiary students from Flinders Street, Adelaide Symphony Orchestra percussionist Bob Hutcheson and emerging professional percussionists such as Ray Horne, James Bailey, Alex Timcke, Vanessa Tomlinson,

Catherine Oates, Tracey Patten, Ian Cleworth, Steven Peterka, Andrew Timcko and Adrienne Sherrin. Although this provided opportunities for young players, a number of whom were also active with their own percussion projects or with Pusz's student ensemble Stix 'n' Tones, a high turnover of players occurred as the majority of these percussionists left Adelaide for further study or jobs elsewhere. Lacking a regular membership base made it difficult for Adelaide Percussions to continue to make projects, and during the 1990s Pusz formed additional percussion ensembles including steel pan and taiko ensembles, and percussion quartets percusSON (1999–2000) and Sonitus. Composed of Vanessa Tomlinson, Catherine Oates, Steven Peterka and Ryszard Pusz, Sonitus presented a number of favourably reviewed concerts in the 1990s featuring a combination of established international repertoire by Cage and Reich with commissioned Australian works. Some indication of the ideals of the group can be found in a Sonitus concert publicity blurb that commands 'stand apart from the slagheap of gutless conformity', a directive met with an enthusiastic audience response (Smith, 1990).

The turn of the century saw the formal end of all Adelaide Percussions activity, and a slowing of adjacent percussion activity when the Flinders Street School of Music amalgamated with the Elder Conservatorium at the University of Adelaide. The conservatorium absorbed the Flinders Street percussion instrument collection, and rehearsal space was no longer available to percussionists outside of the conservatorium student body. Adelaide Percussions was operational for 22 years, and all of the previous core members of Adelaide Percussions are still active performers and pedagogues in Australia. Pusz continued to commission works for solo percussion and mixed chamber ensemble from Australian composers, such as Becky Llewellyn's *The Wretched Hourglass* for marimba and violin (2002) and expatriate Jennifer Fowler's *Towards Release* (2005); and international composers, such as Richard Lavenda's *Split 5 Ways for Marimba* (1992). Works for Pusz's students were commissioned, including Llewellyn's *Early Spring* for glockenspiel or vibraphone (1988) and a collection of six works by Eric Bryce for mallet percussion and piano, however the majority of these works also had limited impact due to single performances.

Experimental percussion in 1990s Adelaide

Although Adelaide Percussions' activity slowed in the 1990s, this decade was a productive time for percussion in Adelaide with activities at the Elder Conservatorium. It was during the 1990s that James Bailey began making marimbas. The design of the Yamaha five-octave marimba in 1984 led to the composition of increasingly virtuosic marimba repertoire by composers in North America, Japan and Europe, and this international repertoire was followed by Bailey and his students. As the expense of freighting five-octave instruments made overseas to Adelaide was prohibitive, Bailey initially began making marimbas to enable performance of this music. This in turn led to increased interest in marimba performance in Adelaide. Bailey's contributions to the Adelaide percussion scene through pedagogy and instrument making were celebrated during the Elder

Conservatorium centenary events in 1998, with a commission from Rob Cossom titled *Concerto Grosso* for nine percussionists using instruments made by Bailey including marimbas, glockenspiel and drum set.

It was also in the 1990s that a burst of experimental percussion activity existed for a brief yet fertile period through the activities of Vanessa Tomlinson and Catherine Oates. Oates reflects that 'the early nineties were full of amazing challenges, incredible momentum, and a feeling of great excitement' (Oates, 2017). In addition to performing with Adelaide Percussions, both Tomlinson and Oates were active as soloists, in their duo Rhythm Method and with the ACME New Music Co. The ACME New Music Co. was a prolific collective of performers and composers engaged in contemporary and experimental music activity in Adelaide. The group presented countless new works for solo and chamber ensembles in concerts, at times giving entire concerts of world premières every week in response to ideas, instruments or influences on group members at the time (Tomlinson, 2017b). Tomlinson reflects that it was during this time that the tam-tam became a central feature of her practice:

> In 1990 I walked into John Reynolds Drum Shop in Adelaide with my drum buddy Hej (Catherine Oates). At the back of the shop, slightly raised, was a 38-inch Paiste tam-tam. I asked the shop owner if I could hit the instrument. I was given a large white fluffy mallet and went up and struck the gong – with a similar style to Red Simmons from *Hey Hey Its Saturday* (for Australian viewers only). Every snare drum in the shop rattled, but I was close enough to the tam-tam to hear it transform as it decayed, to get the tiniest glimpse into its sonic potential. I wanted it. After negotiating to pay it back over the coming year, I left the shop with the tam-tam on the roof of Hej's Mini, she steering with one hand and holding the tam-tam with the other. I changed gears with one hand and held the other side on the roof. We arrived at my house, set up the tam-tam, hit it once and then sat on the couch and listened until the sound totally disappeared. This tam-tam was taken for walks to parks every weekend, as I attempted to play it as loud as it would go, wrapping my ears in scarves, beanies; and the tam-tam still not reaching its limit. With composer David Harris I took the tam-tam to a whispering wall, experimenting with the sound across the concave wall, and through the valley. Composers wrote solos for the instrument – Harris focussing on durational rolling in discreet positions (Harris, 1993), Raymond Chapman Smith on screaming Dada poetry through the instrument picking up specific frequencies (Chapman Smith, 1993), and David Kotlowy on drawing Japanese characters with superballs on the face of the instrument (Kotlowy, 1993). We were all obsessed by the universe of sound inside this amazing instrument.
>
> (Tomlinson, 2017a)

This instrument was the source of inspiration for at least five works for solo tam-tam. The closing sentences of the preceding quote allude to the climate of exploration of sound created by the ACME New Music Co. at the time. A number of concert programs bore the statement: 'Definition of percussion; anything you can bang, scrape or shake. Definition of Harris' music; a shitload of gorgeous little things', further alluding to the inclusive scope of their percussion concerts. Oates describes the creative environment as prolific and exciting: 'we were surrounded by composers who were producing new music for us what felt like daily. They would ask about what instrumentation and form we might be into at that moment, then rock up the next week with a complete new work for us to play' (Oates, 2017). Although the scores or performance instructions for a number 'engaging pieces by Adelaide composers' (Cary, 1993) have since been lost, concert programs show that over 40 percussion works were composed for Tomlinson and Oates as soloists, Rhythm Method or quartets within ACME New Music Co. between 1990 and 1993 alone. Oates describes a memorable event:

> A particular favourite was a concert we did in the Little Theatre in the Adelaide University. As we came onstage to bow when it was all over, I was covered in sweat, utterly spent, the lactic acid that had built up in my arms was making my whole body tremble, I looked around the stage at the detritus left behind – a row of classroom chairs one of us had kicked over, tree branches all over the place, bouncy rubber balls that had bounced everywhere, sticks on the ground, having been dropped or thrown, the flotsam and jetsam aftermath of a Rhythm Method gig – and thinking how lucky I was to be playing at this ridiculously fertile time, and to be surrounded by such excellent and talented people.
>
> (Oates, 2017)

Although this burst of creative activity saw the première of dozens of new works, the scores or performance instructions of the majority of these works have mostly been lost as they were not published. This prolific period was documented by the ABC through the recording of a high number of concerts of Australian music (Tomlinson, 2017b). The experimental nature of the works themselves lent the works to documentation by recording rather than by publishing scores. Many of these were based on the exploration of a single idea by a specific performer in performance and therefore were not fully notated as 'writing a piece then was more like doing an improvisation now. They were all experiments, which is why they didn't last' (Tomlinson, 2017b). One advantage of this style of making work was that is allowed high-volume exploration of new ideas. For example, in 1991 alone Tomlinson and Oates premiered more than 20 new works by Becky Llewelyn, John Polglase, Stephen Whittington, David Kotlowy, Raymond Chapman Smith and David Harris. The titles of a number of works by Chapman Smith and Harris in 1991 demonstrate both the frequency that new works were being created and performed and the thematic link of performer exploration between them. Chapman Smith wrote five works for different instrumentation titled *Untitled 23.2.1991*, *24.2.1991* (both for solo marimba), *Untitled 26.2.91* (timpani duo), *12.11.1991* (marimba solo), and *Untitled 19.10.91* (four marimbas). In the same year, Harris wrote eight percussion works titled *Thrash 14.2.1991, Thrash 11.3.1991*, *10.8.91*, *3.9.1991, 4.9.1991, Anonymous 13.11.90*, *17.11.1991* and *1.12.1991*.

The solo performances given by Tomlinson of works by the six composers mentioned earlier as well as Jason Allen, Jeremy Rowney, Lyndon Northeast and Quentin Grant marked some of the earliest experimental solo percussion activity in Australia. Although a number of solo percussion works by Wesley-Smith, Westlake and others discussed in previous chapters were receiving regular performances by various percussionists, the point of difference in the recitals by Tomlinson and Oates is the proportion of experimental works influenced by the improvisational and exploratory practices of the musicians of ACME New Music Co., and the focus on making art through the medium of music rather than a focus on making music through the medium of concerts. In a 1996 interview Tomlinson stated 'I'd rather be seen as an artist (that is a "maker of art") than a percussionist' and this led to concert presentations influenced by theatre or visual art designed to draw Adelaide audiences into some of the more experimental works (Weekes, 1996). For example, in 1993 Tomlinson utilised a live feed of the percussionist's perspective in performance via 'a large TV display showing the relevant action in close-up' (Cary, 1993). A series of newspaper and magazine articles suggests that these efforts were effective at drawing potentially doubtful audiences 'into the atmospheres and ambiences she generates, showing not the least sign of self-consciousness even when her actions would appear, under normal circumstances, ludicrous' (Silsbury, 1996).

Additionally, ACME New Music Co.'s flexible format saw the co-presentation of concerts with Tomlinson and Oates' duo Rhythm Method, who performed works by leading twentieth-century international composers such as Christian Wolff, Maki Ishii, Harold Budd and Alvin Lucier, and Australian premières by Becky Llewellyn, Stephen Whittington, John Polglase and many others including those mentioned earlier. It was during this time that Tomlinson unknowingly gave the Australian premières of seminal solo percussion works such as Cage's *One4*, Brian Ferneyhough's *Bone Alphabet*, Vinko Globokar's *?Corporel* and Xenakis' *Rebonds*, often shortly after their international premières (Huang, 2017). In 1990s Australia, the primary external source of new music was publishing houses through regular contact with publishers, from whom scores were ordered via post and a number of these premières were the result of discoveries Tomlinson made while communicating with publishers.

In a 1996 interview, Tomlinson alluded that the small contemporary music scene in 1990s Adelaide became a positive influence: 'in Adelaide one has to come up with one's own solutions. There's no hierarchy, no big names, so you have to rely on your own resources and that can be a distinct advantage . . . it's up to the individuals to get on with their own work' (Weekes, 1996). This sentiment was echoed over 20 years later by Fleur Green (Green, 2017).

Adelaide since 2000

Despite a flourishing scene in percussion education and a number of individual artists such as classical vibraphone soloist Nick Parnell and alternative pop vibraphonist-vocalist Fleur Green, no ongoing professional contemporary percussion ensemble has emerged in South Australia since 2000. Johnson speculates that this is partially a result of the size of Adelaide and the amalgamation between Flinders Street and the conservatorium:

> Everyone at Flinders Street was encouraged to get out into the community and play, but at the Conservatorium they were expected to play in gigs at the Conservatorium. There is no other collection of gear and this is possibly the reason that no other group has been formed since.
>
> (Johnson, 2013)

While this may or may not be the case, the increased ease of importing percussion instruments and comparatively reduced costs suggests that this is likely to be only one of many contributing factors. Today, it is common for individual percussionists to purchase their own instrument collections over many years, and to pool these instruments in the formation of a percussion ensemble. Grigg also notes the impact of the amalgamation of tertiary music institutions on Adelaide percussion activity, as the reduction from three institutions to one has resulted in less support for emerging percussionists in the form of access to crucial infrastructure such as rehearsal studio space and venues equipped with percussion instruments (Grigg, 2017). Green observes that this factor, coupled with rigorous technical training in Adelaide, prepares percussionists for freelance careers with major performing arts organisations, and that the trend of leaving Adelaide to pursue work overseas has continued in the twenty-first century with a number of Adelaide-trained percussionists securing positions with Canadian organisation Cirque du Soleil and various theatre shows in Las Vegas (Green, 2017). Regardless of the reasons, a clear niche in the percussive arts space exists in South Australia at present, waiting for a new generation of percussionists to step forward and continue expanding Australian contemporary and experimental percussion practices.

Note

1 This date has been provided by Middenway via email communication with the composer. Middenway has composed a number of works titled *Mosaics*, and describes them as a 'set of variations'.

References

Cary, T. 1993. "Music: Tristram Cary in Adelaide." *The Australian*, 9 July.

Chapman-Smith, R. 1993. *Reply the Night* [score]. Composer's manuscript.

Harris, D. 1993. *Tam Tam 12.8.93* [score]. Composer's manuscript.

Huang, A. 2017. "Examining Australian Contemporary Percussion through 'things Uniquely Australian': Interviews with Louise Devenish, Claire Edwardes, and Vanessa Tomlinson." Contemporary *Music Review*, 36 (1–2): 102–117. doi:10.1080/07494467.2017.1368175

Kotlowy, D. 1993. *Untitled* [score]. Composer's manuscript.

Martin, P. J. 1996. "Synergy Percussion: Australian Percussion Music in Context." Master of Music thesis, Armidale: University of New England.
Silsbury, E. 1996. "A Performance that Goes beyond Music." *The Advertiser*, 30 July.
Smith, R. 1990. "Skilful Sounds Tailored to Surroundings." *The Advertiser*, 25 August.
Smith, R. 2003. "How the Balloon Taught the Piano to Play." *RealTime*, 55: 28.
Tomlinson, V. 2017a. "On Listening: A Universe of Sound." C*ontemporary Music Review*, 36 (1–2): 5–14. doi:10.1080/07494467.2017.1368177
Weekes, D. 1996. "Big Noise, Quiet City." *Adelaide Review*, circa 1996. Accessed personal archives of Vanessa Tomlinson.

Personal communication

Green, F. 2017. Telephone Interview with Louise Devenish, 15 November. Adelaide, South Australia.
Grigg, A. 2017. Telephone Interview with Louise Devenish, 23 November. Adelaide, South Australia.
Johnson, D. 2013. Telephone Interview with Louise Devenish, 28 November. Adelaide, South Australia.
Johnson, D., and R. Pusz. 2014. Live Conversation with Louise Devenish, 6 March. Adelaide, South Australia.
Lloyd, R. 2013. Email Communication with Louise Devenish, 17 December. Adelaide, South Australia.
Oates, C. 2017. Email Communication with Louise Devenish, 2 December. Adelaide, South Australia.
Pusz, R. 2013. Skype Interview with Louise Devenish, 13 May. Adelaide, South Australia.
Pusz, R. 2014. Email Communication with Louise Devenish, 27 March.
Timcke, A. 2013. Live Interview with Louise Devenish, 24 July. Perth, Western Australia.
Tomlinson, V. 2014. Telephone Interview with Louise Devenish, 5 July. Brisbane, Queensland.
Tomlinson, V. 2017b. Personal Communication with Louise Devenish, 10 November.

5 Percussion in Western Australia

As the most isolated capital city in the southern hemisphere, in the 1960s and 1970s Perth was disconnected from contemporary percussion developments taking place in other parts of Australia. There was some percussion activity in orchestras and brass bands, however as few trained percussionists lived in Perth, percussion in these ensembles was usually played by other instrumentalists. Consequently, when international artists such as Les Percussions de Strasbourg and George Gaber included Perth on their Australian touring itineraries in 1971/1976 and 1975, respectively, the impact they had on the local community was limited. When the same artists performed in New South Wales, South Australia or Victoria, their audiences included conservatorium-trained and professional percussionists, who had both the skills and resources to subsequently begin exploring contemporary percussion. That there was no similar effect in Perth can be attributed to the fact there was no community of professionally trained percussionists at the time with whom these international artists could engage, further affected by the isolation of the city itself. As a result, the emergence of contemporary percussionists and ensembles in Perth took a slightly different route to that of their colleagues in the eastern states. In order to contextualise this discussion, a brief outline of the contemporary music climate in Western Australia in the 1950s and 1960s will precede examination of contemporary percussion activity between 1970 and 2010.

Prior to the emergence of Perth-based percussion ensembles in the late 1970s and 1980s, one of the few examples of percussion performance outside of the orchestra can be found in the novelty performances of musician, inventor and taxi driver Roy Menzel (1909–2000). Menzel played a number of instruments including banjo, organ, double bass and glass armonica, but it was his novelty percussion performances that garnered the most attention from audiences and local media. Menzel's percussion performances featured chromatic sets of instruments and found objects, shown in Figure 5.1, such as brake drums, hand bells, rubbed wine glasses, circular saw blades (which were spun on a sheet of marble to produce a clear pitch), klaxon horns mounted on a custom-built frame, and chamber pots struck with mallets upon which he performed popular tunes. In addition to building frames to arrange these objects for performance and instruments, Menzel also built amusement park rides and held numerous patents for his inventions including rumble strips, the Menzel Hull, the Menzel Air Pump and skateboard

Figure 5.1 Roy Menzel 'Novelty Musician', c. 1963.

Source: Image supplied by Cherie Cadd.

brakes. The invention that had the greatest impact was the Menzel Air Pump, invented in 1939 and purchased by the US government for military use. Menzel's performances came to the attention of the ABC in Sydney, who engaged him to record material for television broadcast on two occasions, in 1962 and 1966. Like It's Percussion, Menzel featured percussion as a solo instrument, albeit playing popular and dance melodies rather than in the creation of new music specifically for those instruments and objects. Thus, his recordings represent a precursor to contemporary percussion activity in a similar manner to Adelaide's It's Percussion (discussed in Chapter One).

The University of Western Australia Conservatorium of Music

The climate for arts funding in Western Australia was strong in the 1950s, and in 1953, two major arts bodies were established at the University of Western Australia (UWA) that provided a stimulus to contemporary music and signalled the beginning of change in the way the performing arts were perceived in the state: Festival of Perth (now Perth Festival) and the UWA Music Department (now the UWA Conservatorium of Music). The UWA Conservatorium of Music was led by Sir Frank Callaway, himself a driving force in music and music education whose roles with the International Society for Music Education and UNESCO's International Music Council gained him an international reputation. Callaway's leadership of a dedicated staff base coupled with new resources enabled the UWA Conservatorium of Music to undertake progressive research in all forms of music, including Australian music and contemporary music. Many UWA staff had worked or studied overseas and had made contact with pioneering Europe and North American thinkers in contemporary music. For example, Callaway met Percy Grainger during his time at the Royal Academy in London; David Tunley studied composition in Paris with Nadia Boulanger; and Roger Smalley, a prominent composer in the European music scene in his own right, attended summer schools with Pierre Boulez and Karlheinz Stockhausen. Like a number of their colleagues in the eastern states, Callaway, Tunley and Smalley returned or migrated to Perth ready to share new ideas with their peers through the composition and performance of new works and to produce new work influenced by their experiences. The maintenance and development of their international professional networks were necessary to sustain and build innovation at the international level they desired; however, there was no substantive platform to support these networks and the dialogues they might foster.

In an effort to rectify the lack of opportunity for international collaboration, the UWA Music Society embarked on a project with Musica Viva to establish an Artist in Residence program. Between 1968 and 1975 this program brought some of the world's leading chamber musicians and composers to Perth, including artists who had made considerable contributions to the development of contemporary percussion globally. Many of the participants of the Artist in Residence program also visited institutions in Melbourne, Sydney and Canberra with the support of Musica Viva, including the aforementioned Les Percussions De Strasbourg and George Gaber. Perhaps the most significant of these visitors to UWA was German composer Karlheinz Stockhausen. Stockhausen frequently featured percussion in his compositions and made major contributions to the international percussion repertoire in *Kontakte* (1959) for piano, percussion and electronics, *Zyklus Nr. 9* (1959), the first solo percussion work composed specifically for a multiple percussion set-up, and *Mikrophonie I* (1964) for tam-tam, microphones and filters. Stockhausen directed a concert of his own compositions at UWA's Octagon Theatre in 1970 before 'an interested but somewhat sceptical audience' (Meyer, 1999: 34). Although no record of the repertoire featured in Stockhausen's Perth concert has been found, it can be assumed that neither *Zyklus Nr. 9*, *Kontakte* nor *Mikrophonie I* were among the works performed and therefore Stockhausen's visit did not serve to introduce live contemporary percussion performance in Perth. This assumption is based on three factors: first, because there were no local percussionists able to perform it; second, a collection of the instruments required for those works did not exist; and third, because the Australian première dates of these pieces have been shown in later concert programs. For example, *Zyklus Nr. 9* was premièred by John Seal at an Australian Percussion Ensemble concert in 1974. *Mikrophonie I* did eventually receive its Australian première in Perth, however this took place in 2013 at the International Computer Music Conference by Speak Percussion, 33 years after Stockhausen's visit to UWA.

In addition to engaging with international artists, an emphasis on the performance and study of Australian music was central to UWA activities at the time, and the Conservatorium of Music engaged in a second collaboration in 1968, this time with the ABC. The Australian Composers' Workshops ran between 1968 and 1975 and they enabled the works of nearly 100 Australian composers to be performed and broadcast in WA. Although the focus was on neoclassical western art music, a small amount of contemporary music by Margaret Sutherland, Nigel Butterley and Barry Conyngham that included percussion in its instrumentation appeared in the workshops' programming. A single percussion work featured in this workshop series, namely, Peter Clinch's *Fugue for Percussion* (1970). This work is the earliest example of a Western Australian composition for solo percussionist and was first performed at one of these workshops in 1971 (Meyer, 1999: 158). Some similarities between Clinch's *Fugue for Percussion* and the early North American percussion repertoire can be found. Like John Cage's early percussion works that focus on the sound of the instruments rather than the performers' technical virtuosity, Clinch's *Fugue for Percussion* does not contain any technical elements such as buzz rolls or decoration. No formal record of who performed *Fugue for Percussion* exists, but one speculates that Clinch, who was a clarinettist and saxophonist, was unable to find a performer trained in contemporary percussion in Western Australia at the time.

Eight years would pass between the premiere of *Fugue for Percussion* and a second new percussion work in WA, namely Paul Sarcich's *2 Side Drum Solos* (1979). Although both the Australian Composers' Workshops and the Artists in Residence programs ceased in 1975, the trickle of international artists through Western Australia continued. Aided by the increasing ease and speed of intercontinental travel and developments in media technologies in the 1970s and 1980s, travel to isolated centres such as Perth began to increase towards 1980, for events curated by Perth Festival and Musica Viva.

The Paul Sarcich Percussion Ensemble (1977–1980)

Prior to his appointment as principal timpanist of the West Australian Symphony Orchestra (WASO) in 1976, New Zealander Paul Sarcich was an active percussionist in Melbourne. Although his primary performance background grew from brass bands in New Zealand and later the Australian Army Band, Sarcich was also a member of the Australian Percussion Ensemble and thus had some performance experience in contemporary percussion. Sarcich sought similar contemporary performance opportunities in Perth and this led to the presentation of a small number of percussion ensemble and solo performances in concert series facilitated by the WASO and the ABC. Although the series presented the opportunity to showcase percussion instruments as solo instruments, it did not afford players the opportunity to program anything experimental, and performances tended towards exhibiting percussion instruments in either a humourous or novelty capacity at a time when this style of percussion performance was diminishing in other Australian cities. For example, Sarcich was approached to give numerous performances of Adolph Schreiner's *The Worried Drummer* (1970) in the WASO's Family

Concert series. The Family Concert series designed to attract families and young people to classical music and Schreiner's theatrical percussion solo focussed on the comedic and novelty aspects of the performance of percussion instruments. Together with harpist Tony Maydwell, Sarcich also frequently performed works for marimba in this series, and believes that this concert series may have marked the first occasion Perth audiences were exposed to the marimba in a solo context (Sarcich, 2012). Maydwell remembers the challenge of sourcing repertoire for this instrumentation: 'We didn't have or know any suitable repertoire so I started looking at some of my other repertoire and ended by transcribing Jacques Ibert's *Entr'acte* (1935) originally for flute and harp' (Maydwell, 2012). In 1977, Sarcich brought his students together as the Paul Sarcich Percussion Ensemble to form part of an ABC Free Sunday Recitals Series program. However, as the repertoire performed by Sarcich and his students at ABC and WASO concert series consisted largely of arrangements of classical and Spanish music, this too is considered as identifying with a classical tradition.

In 1977, the same year as the Paul Sarcich Percussion Ensemble performances, Sarcich travelled to Los Angeles to study with marimbist Karen Ervin. Upon his return, Sarcich was employed to teach percussion to two students at UWA; Felicia Lane and Christine Turpin. Sarcich owned the only marimba and vibraphone in Perth at the time and he used these instruments to teach the new Leigh Howard Stevens four-mallet technique to both students, a technique new to Australia at the time. Unfortunately, a lack of opportunity and like-minded colleagues led to the departure of all of three percussionists from Western Australia within one year. Both Turpin and Lane pursued orchestral performance: Turpin transferred to the Victorian College of the Arts, and Lane moved to London and enrolled at the Royal Academy of Music under famed percussion historian James Blades. In 1980, Sarcich migrated to London to develop his career as a composer.

Nova Ensemble (est. 1983)

In 1981, VCA graduate David Pye replaced Sarcich as principal timpanist for the WASO. Besides principal percussionist Brian Sadique, the WASO percussion section was composed of enthusiastic but untrained casual players at this time, including the orchestral librarian, a nurse and a French horn player. Pye recognised the need to begin training young percussionists for the WASO section and began teaching his first pupils Amanda Dean and Paul Tanner in 1983. This training subsequently became the foundation upon which a substantive contemporary percussion community was formed in Perth.

Pye's initial motivation for teaching percussion had been to lay a professional pathway for students into the orchestra, and individual tuition and orchestral percussion were the primary focus for two years. However, as Pye taught from his home using his personal instrument collection, he soon found that his teaching opportunities were somewhat limited by the instruments available. Thus, Pye soon began purchasing more instruments, with an emphasis on those outside the standard orchestral section, including a 4.3 octave marimba. As students also began to purchase xylophones and marimbas from overseas suppliers there was an influx of percussion instruments to Perth. In 1983 Pye formed the West Australian Percussion Ensemble to create more challenging performance opportunities for his students and a secondary performance outlet for himself. Titled Cycles and Lines, their inaugural program included Henry Cowell's *Ostinato Pianissimo*, Terry Reilly's *In C*, Alan Hovhaness' *October Mountain* and Warwick Bone's *Cycles and Lines*. This was the first program of contemporary percussion music produced in WA and this was recognised in the form of a national live broadcast by the ABC. The program was performed by percussionists Amanda Dean, Mandy Pummer, David Pye, Paul Tanner, Christine Turpin, David Vrcic and Michael Watson with ten guest instrumentalists for *In C*: Mary Johnston, Graham Simpson, Joan Wright, Wendy Clarke, Leanne Glover, Philip Tornatora, Brett Fitzsimmons, Kim Harrison, Robert Braham and Craig Ogden.

Continuing the national trend of programming significant works from the international repertory in combination with new Australian music was vital in the establishment of the WA Percussion Ensemble's identity. *Ostinato Pianissimo, October Mountain* and *In C* were established international percussion ensemble works, and flute and percussion duo *Cycles and Lines* by Warwick Bone was composed for Pye when he was a student at the VCA. With the exception of Christine Turpin, at the time WASO's acting principal percussionist, all of the percussionists in the WA Percussion Ensemble's first concert were Pye's students. As there were not enough trained percussionists in Perth to play repertoire that was both readily available and suitable for their skill level, the program also included works involving other instrumentalists. Some of these instrumentalists also played percussion in this performance.

Following the positive reception of their first performance, Pye decided to continue to create contemporary music concerts, and this marked the beginning of a prolific period of percussion activity in Perth. In 1983, Pye discarded the name WA Percussion Ensemble in favour of Nova Ensemble to reflect his passion for conducting and interest in collaboration with other instrumentalists, particularly the repertoire for two pianos and two percussion. The previous name of WA Percussion Ensemble suggested a purely percussion ensemble; however, Nova Ensemble was designed to be a mixed chamber ensemble with 'percussion at its heart' (Pye, 2012). This gave the group a flexible form which allowed them to present major classical works of the twentieth century for piano and percussion in one season, followed by a concert of original music for chamber orchestra or a program for percussion quartet in the next season. The regular inclusion of percussion instruments in concerts was new to Perth audiences and many came to Nova Ensemble concerts to experience the unique and visual aspects of percussion performance. Despite Nova Ensemble's efforts to be known as a contemporary chamber ensemble, the strong percussive elements of their programs and the fact that their director was a percussionist inevitably saw the group become known as a percussion ensemble. Additionally, a series of reviews and media campaigns frequently described the group as a percussion ensemble and it was a categorisation that stuck. Pye states:

> We have done programs of percussion music but we're not a percussion ensemble, we never have been. There is a heavy percussion influence, I'm a percussionist, but I don't know how many times I've had to say, 'We are not a percussion ensemble' . . . and then presented a program of four percussionists. But the number of non-percussion concerts we've done outnumbers the only-percussion concerts by far.
>
> (Pye, 2012)

Regardless of how they preferred to be perceived at the time, Nova Ensemble's production of contemporary music concerts that featured significant international percussion repertoire was pivotal in raising awareness of contemporary percussion in Perth. Nova Ensemble's performances of Iannis Xenakis' *Persephassa* in 1987 moved percussion to the forefront of contemporary music in Western Australia, and the ensemble further cemented this position with the production of the interdisciplinary project Mirror Coda. Created for the 1989 Festival of Perth, Mirror Coda brought Nova Ensemble together with the Chrissie Parrott Dance Collective, a Perth-based contemporary dance company directed by its namesake. Performed over four nights in February 1989 at the Quarry Amphitheatre, Mirror Coda was the first major collaboration between contemporary dancers and contemporary percussionists and pianists in Australia (premièred three years before Synergy with Synergy). The Quarry Amphitheatre is an outdoor venue, making it perfectly suited for a program that included George Crumb's *Makrokosmos III: Music for a Summer Evening* alongside Steve Reich's *Sextet*, Xenakis' *Psappha*, Peter Hadley's *Mimic Time* and Roger Smalley's *Percussion Quartet*. Figure 5.2 shows a rehearsal for this program. These works were performed to large audiences by members of Nova Ensemble in the centre of the stage, surrounded by dancers from the Chrissie Parrott Dance Collective. These artists continued to collaborate in the 1990s, on projects such as Terminal Velocity (1991) and Terra (1992).

The artistic risks taken by Pye when programming ambitious repertoire to an untested audience proved to be worth the effort when capacity audiences and reviewers consistently received their programs positively. Described by *The Australian* newspaper's music critic Neville Cohn as 'astonishing and exciting' (Cohn, 1987), Cohn encouraged Perth audiences to seek out this music with statements such as 'one hopes profoundly this first performance is the precursor of many more' (Cohn, 1986). Pye describes this response by asserting:

> The simple fact was, there was no-one else in town playing repertoire like that. So we did *Persephassa, Pléiädes, Tehillim, Sextet, Makrokomos III, Ionisation*, the Carl Vine quartet [*Defying Gravity*, 1987] and so on. We did quite a lot of Steve Reich and not just his percussion works. And the thing was they were the first performances [in Western Australia]. None of us had ever played that stuff before and people here had never heard it before. So inevitably it had to have an impact.
>
> (Pye, 2012)

Figure 5.2 (L-R) Nova Ensemble's David Pye, Amanda Dean and Paul Tanner rehearsing Steve Reich's *Sextet.*

Source: Photograph John Austin 1987.

Although Nova Ensemble continued to perform major international contemporary repertoire, they began to include works composed by members of the group, as the APE and Synergy Percussion had done before them. This became the central focus on Nova Ensemble's programming within a few years, leading to entire programs of works composed 'in-house'. The artists of Nova Ensemble began collaborating with other local musicians who specialised in electronics, various global traditional musics and instrument building. These collaborations fundamentally influenced Nova Ensemble's sound and style and the repertoire performed by the group began to take a new direction. Pye began collaborating with composer-performers Cathie Travers (accordion, electronics and piano), Lee Buddle (winds) and Ron Reeves (percussion) and the individual voices of these composers began to assert their influence on the Nova Ensemble sound. Although all of these artists originally came from western art music backgrounds in Australia, each had studied or explored various traditional musics from around the world. The influences of these studies began to filter into Nova concerts and a collective interest in the rhythms of neighbouring countries became clear. In the beginning, this was led by Ron Reeves' extensive study of the music of Indonesia and India.

A multi-instrumentalist with an interest in composition and instrument building, Reeves was also one of the founding members of The Sydney Percussion. However, in 1976 Reeves left Sydney to teach western percussion in Indonesia. During the five and a half years he spent in Indonesia, Reeves developed a keen interest in traditional Indonesian music. He studied a number of traditional instruments in different parts of Indonesia, including various instruments used in gamelan ensemble music and the taganing, a set of five tuned drums used by the Batak Toba tribe in North Sumatra. Upon his return to Australia, Reeves brought a unique sound to Nova's projects via the introduction of Indonesian instruments and playing techniques in a number of original compositions influenced by the traditional music he had studied. Pye refers to a group composition project in 1986 that involved the use of Reeves' Indonesian influenced music as the genesis of this Nova 'sound':

> We worked for a week in a church in Subiaco and created a program that we performed at the end of the week. That was all original stuff and looking back, what I wrote was an appalling piece of music! But Lee wrote *Just An Inkling for an Angklung* then. And we got into Ron's music doing vocal percussion and gamelan stuff and Cathie had her *Antarctic Dreaming*. That concert was the genesis of the Nova of the '90s because it led us into the whole area of improvisation, electronics, group-composed works . . . composition by the performers rather than compositions by 'composers'. So that was a pivotal week that week and things took off after that.
>
> (Pye, 2012)

This success of this project saw this multicultural approach to percussion ensemble music develop into a significant part of Western Australian percussion practices in the 1990s, parallel to similar approaches in Sydney during the same period. Although a small number of Indonesian instruments owned by Reeves were available to Nova Ensemble, other traditional instruments were not always available. In other cases, a different sound altogether was required. When members of Nova Ensemble became interested in instruments from Asia and India and began to explore the possibility of acquiring more exotic instruments they discovered high costs, extended delivery timeframes and the necessity of purchasing instruments sight unseen problematic. Additionally, stringent border protection authorities could not guarantee instruments would be allowed to pass intact through customs. When this occurred, percussionists made use of everyday items (also referred to as found objects) as instruments and/or built new instruments out of easily accessible natural and manufactured materials, a practice first explored in depth by Harry Partch and Lou Harrison in the 1940s (Partch, 1974; Amirkhanian, 1989). The instruments used in early North American percussion ensemble repertoire featuring found objects served as a point of departure for Nova Ensemble, who began to explore the possibilities of making instruments that were designed to produce similar sounds to Reeves' authentic traditional instruments using materials at hand. Building their own instruments proved a wise decision for another reason. Many of Nova Ensemble's early works included Indonesian angklung and gamelan instruments from Reeves' collection, and his departure from the group demanded alternative means of producing these sounds, such as making substitutes. Instrument building subsequently became another major part of Nova Ensemble's practice in the 1980s and 1990s. Additionally, offshoot ensembles from Nova Ensemble were also created that relied on the specific skills and instruments of the performers, such as Indian Ocean Unit (IOU), a trio comprising Reeves on mrdangam and kendang, Pye on marimba and Buddle on winds.

A number of instruments built and designed by Nova Ensemble were constructed for the group's 1994 Junkelan project. As the name suggests, Junkelan took musical cues from Indonesian gamelan and combined these with digital music technology. All of the instruments were designed and built by the ensemble using scrap metal and industrial plastics, and the cog bonang were one such instrument. The cog bonang were modelled after an instrument commonly found in Javanese gamelan. Javanese bonang consist of two rows of small, pitched bossed gongs, and are tuned to either the pelog or slendro scales. The Nova Ensemble's cog bonang is shown in Figure 5.3. The cog bonang was made of discarded metal cogs from machinery arranged in pitch order, however they were not selected with the intention to form a particular scale. Like Lou Harrison's American Gamelan, although the cog bonang bore resemblances to Javanese gamelan instruments, it was used to play contemporary western music rather than Javanese-influenced music. All of the other instruments made for Junkelan (such as the circular sawrus shown in Figure 5.4) were based around discarded scrap metal and are listed in Table 5.1.

The building of instruments using found objects provided many opportunities for exploration, and Nova Ensemble created a number of shows in this manner, including Boxes (1991), Trash (1996) and Pockets (2000). All three of these productions used instruments created from recycled materials and the music itself was 'informed by our proximity to the neighbouring Indian Ocean region, yet

Figure 5.3 Nova Ensemble's David Pye, Lee Buddle and Paul Tanner playing the cog bonang.

Source: Photograph supplied by David Pye.

Figure 5.4 Nova Ensemble's circular sawrus.

Source: Photograph supplied by David Pye.

[was] unmistakably Australian' (Cain, 2013). Additionally, instrument building and interest in the music of the countries of the Pacific Rim led Nova Ensemble into collaborations with AC/PVC, an ensemble also based in Fremantle. Indeed, the two groups shared a common membership in Mark Cain and Ron Reeves and thus influenced one another's creative work in the late 1980s.

AC/PVC (1987–1990)

Founded in 1987 by Nova Ensemble's wind instrumentalist Mark Cain and pan-piper Peter Keelan, AC/PVC straddled the boundaries between contemporary percussion, instrument building and world music, and another type of percussion performance emerged in Western Australia through their activities. AC/PVC performed exclusively on junk percussion and wind instruments they designed and built themselves out of Poly-Vinyl-Chloride (PVC) piping, directed by Keelan. Keelan is an Andean music specialist who studied instrument making in Chile, and this knowledge was applied to the junk percussion instruments built by members of the group, which included Nova Ensemble member Ron Reeves and Indian musician Kuppusany Raman, shown in Figure 5.5, as well as Kerry Fletcher and Aden d'Adhemar, Paul Tanner, Chris Waite and Gary Ridge.

Table 5.1 Percussion instruments made by Nova Ensemble for Junkelan

Nova Ensemble instrument	*Description*	*Inspiration*
Cog bonang	Scrap metal cogs and circular component of machines formed the gongs tuned to an approximated Eb major scale. These were welded at a single point to small metal upright pegs on a rectangular metal frame.	Javanese bonang
Stopsignaphone	Rectangular metal 4 x 2 inch signpost tubing arranged to form an indeterminately tuned mallet instrument, similar to the sixxens designed by Iannis Xenakis for the sextet *Pléiädes*.	Sixxens
Circular sawrus	A collection of different sized circular saw blades mounted like suspended cymbals and struck with mallets to produce a ringing tone. See Figure 5.4.	Mallet percussion
Oil drum taiko	Two 44-gallon drums bolted together. The metal base ends were replaced with synthetic drum heads to create a two-headed drum.	Odaiko
Chain maker	A heavy metal chain dropped into a 44-gallon drum raised slightly off the floor to make a loud clanging sound.	Tam-tam
Heckle-and-jeckle	A metal spring bolted at both ends to an approximately 30-centimetre long piece of rectangular metal 4 x 2 inch signpost tubing. The spring was scraped with a thin metal rod.	Guiro

This overlap in members resulted in the gradual filtering of instrument making and junk percussion into Nova Ensemble repertoire. Junk percussion music refers to an energetic, groove based form of percussion that often uses rhythms inspired by world music and is played on instruments made from found objects. Junk percussion is popular around the world, with numerous ensembles and music publishers making regular contributions to the genre such as Scrap Arts Music (Canada), Blue Man Group (USA), STOMP (UK) and Row-Loff Productions (USA). Its popularity can be attributed to the huge variety of timbral qualities available, its affordability and the distinctiveness handmade junk percussion instruments can provide an ensemble. A 1987 AC/PVC concert program describes 'the basis of their instrument collection as percussive' and lists zamporias (panpipes), flutes, didgeridoo, tube drums, thong-a-phones, an A-frame marimba, talking drums and a variety of small hand percussion instruments in their collection. As this list suggests, drawing on instruments from countries as diverse as Chile, Indonesia, Scotland and Africa for inspiration on their PVC instrument design was a common practice for the group. The use of humour and novelty ideas was also prioritised,

Figure 5.5 AC/PVC members Mark Cain with talking drum, Kuppusany Raman in front of A-frame marimba and Peter Keelan leaning over a thong-o-phone.

Source: Photograph supplied by Louise Devenish.

and is most evident in instruments such as the didjeriloo, surgical glove bagpipe (see Table 5.2) and an enormous flute that required one player per note to perform.

In addition to designing and building all of their own instruments, each member of AC/PVC composed music for the ensemble to perform on their unique instrument collection, and each drew on various world music cultures and contemporary music practices for inspiration. Each member of the group had studied various indigenous forms of music around the world, and this contributed to the distinctly multicultural sound of AC/PVC's music. As with Nova Ensemble, music from Indonesia and other countries around the Pacific Rim provided the greatest source of inspiration. AC/PVC and Nova Ensemble regularly produced concerts together, juxtaposing substantive international and Australian percussion music with junk percussion repertoire to great effect. Collaboration was a key stimulus and both groups collaborated with other instrumentalists including accordionist Ross Bolleter and contemporary dancers Jean Tally and Claudia Alessi. These collaborations raised AC/PVC from novelty entertainment to a legitimate musical entity while retaining the sense of fun intrinsic to their work. When the ensemble disbanded in 1990, AC/PVC left behind a collection of their other percussion instruments listed in Table 5.2, including a thong-a-phone that was absorbed first

Table 5.2 AC/PVC instruments (Note that the names thong-a-phones and thong-o-phone were used interchangeably by AC/PVC.)

Instrument	*Description*	*Inspiration*
A-frame marimba	A chromatic idiophone with PVC resonators played with mallets by two players, one on each side of the instrument	Marimba
Talking drum	AC/PVC's version of a talking drum used two different sizes of PVC piping for the body of the instrument	West African talking drum
Thong-a-phone	A chromatic idiophone consisting of a row of PVC pipes whose open ends are struck with rubber thongs to produce a low sound similar to a bass marimba	Bass marimba
Octothongo	A bass thong-a-phone with a range of one chromatic octave	Bass marimba
Zamporias	Pan-pipes made of PVC piping	Andean panflutes
Tube drums	Drums made from a single size of PVC piping	Various drums
Thongchen	An alto thong-o-phone on a triangular frame with a range of two chromatic octaves	Marimba
Customised octobans	PVC pipe, rope and garden hose piping drums fitted with plastic drum heads	Octobans or boobams
Bombos	Drums made from Acrylonitrile Butadiene Styrene (ABS) welded ducting	Various drums
PKdrum	Made from curved ABS ducting and performed by Peter Keelan	Various drums
Rondrum	Made from Acrylonitrile Butadiene Styrene (ABS) welded air-conditioner ducting for Ron Reeves	Various drums
Sir Rondrum	A larger version of the Rondrum that stood approximately 2 metres high	Rondrum
Surgical glove bagpipe	Made from two plastic water bottles, a surgical glove and three plastic tubes. Two of the tubes were attached to the water bottles and served as drone pipes, while the third had finger holes similar to a whistle or recorder.	Bagpipes
Fufu pipe	Small wind instruments made of balloon rubber stretched tightly over a film canister to create a vibrating membrane. The canister is fixed to a pipe with finger holes. Air is blown into the canister through a second small pipe that functions as a mouthpiece.	Various wind instruments

(*Continued*)

Table 5.2 (Continued)

Instrument	*Description*	*Inspiration*
Megabass clarinet	A bass clarinet mouthpiece attached to wide PVC piping to make a quasi bass clarinet with a range of one octave	Bass clarinet
Didjeriloo	A toilet bowl and piping adapted to be played in a similar manner to the didjeridu	Didjeridu

into the Nova Ensemble collection and then used by percussionists Paul Tanner and Amanda Dean in world music ensemble Ozmosis. Cain continues to produce projects using PVC wind instruments and found objects as an independent artist (Ryan, 2016).

A lack of access to vast instrument collections has had a significant impact on the work produced by other ensembles and individuals around Australia. The instrumentation of many twentieth- and twenty-first-century Australian works is unique in accordance with what was available to percussionists at the time of composition. For example, Michael Askill's interest in gongs from around the world (Askill, 2017) is reflected both in Synergy Percussion's numerous catalogue of works featuring gongs, and in more recent repertoire for Early Warning System. Similarly, Vanessa Tomlinson's *Mutant Theatre* commissions from Anthony Pateras were scored for found objects in part because she had sold her marimba and vibraphone to finance overseas study in the 1990s. Graeme Leak's extensive investigations into instrument building drew on the hardware stores nearest his home.

The Western Australian Academy of Performing Arts

The 1980s was a time of prosperity for Western Australia in general, as the world watched host city Fremantle in anticipation of the 1987 America's Cup. It was a time of growth for arts and culture, and this included the establishment of the Western Australian Academy of Performing Arts (WAAPA) in 1980. In 1987 Richard Gill, then the dean of the WA Conservatorium of Music at WAAPA, engaged American percussionist Gary France to establish WA's first formal percussion course. Although percussionist Paul Tanner studied at the University of Western Australia prior to this, it was on a single study basis. Tanner states: 'The university had owned a small collection of [percussion] instruments before I got there, for the orchestra, but they sold it the year I arrived. I was the only [percussion] student at the time and had to bring in my own instruments for every performance at the university' (Tanner, 2012). Thus, France's appointment represented the first opportunity for percussionists to study in a specialist department at a tertiary level in Western Australia. France's study at the Crane School of Music in New York and the University of North Texas encompassed what France refers to as 'total percussion' (France, 2014), defined as a broad practice across various percussion instruments and a range of western musical genres from orchestral to jazz. France implemented a percussion course integrating drum set and percussion at WAAPA based on his experiences as a student in North America. The WA Conservatorium of Music Percussion Ensemble (WACMPE) was formed, and France also became involved with Nova Ensemble and the percussionists of the WA Symphony Orchestra.

Notably, WAAPA purchased a large collection of percussion instruments including those outside of the standard orchestral instruments such as a 4.6 octave marimba, a bass marimba, and two chromatic octaves of flat gongs. In addition to enabling students to explore a wider range of percussive sounds and techniques, this collection of instruments meant that other Australian and international contemporary percussionists could begin to visit Perth on a more regular basis to interact with and provide additional education to students. In the 1970s, when international and interstate visitors including Les Percussions de Strasbourg and the Australian Percussion Ensemble performed in Perth, they needed the support of large organisations such as the Perth Festival to cover the cost of freighting their instruments. Although it came almost 20 years after the introduction of percussion to tertiary programs in South Australia, the establishment of WAAPA's percussion course in combination with the work of Nova Ensemble and AC/PVC represented an opportunity for further development of contemporary percussion in Perth.

A community began to form around the activities at WAAPA, and the first year of operation coincided with percussionist Graeme Leak's solo Australian tour. Gill invited Leak to be artist in residence at WAAPA for one week, and during this time Leak worked with members of WACMPE, Nova Ensemble and the wider percussion community on the WA première of Varèse's *Ionisation*. When France left WAAPA in 1994 to establish a similar course at the Australian National University, he was replaced by Chris Tarr for drum set tuition and Tim White for percussion tuition. White renamed the student percussion ensemble Defying Gravity, after Carl Vine's quartet work of the same name, and expanded the ensemble to include his students from the University of Western Australia. By this time, percussion in Western Australia was growing, and in 1996 a second professional percussion ensemble emerged: Tetrafide Percussion.

Tetrafide Percussion (1996–2014)

In 1996 Iain Robbie, Robyn Gray (nee Sarti), Jennifer Winley and Neville Talbot formed Tetrafide Percussion. Although the group's name stemmed from the prefix 'tetra', meaning four, from their second performance there were five players in the ensemble with the addition of Steven Richter. As with Nova Ensemble, a sense of place influenced Tetrafide Percussion's creative output, and a combination of contemporary percussion and musical influences from countries around the Pacific Rim were evident in the group's style. The most pervasive of these was Papua New Guinean drumming. Iain Robbie had studied percussion as a secondary student in Darwin, during which time he was exposed to traditional Papua

New Guinean (PNG) drumming via Darwin-based group Drum Drum. Following a study trip to PNG with Drum Drum in 1995, Robbie returned to Perth with a collection of garamuts (PNG log drums), kuakumbas (bamboo flutes played in pairs) and permission to share the traditional rhythms he had learnt with Drum Drum. Subsequently, Robbie arranged some of these rhythms for Tetrafide into a work known as *Papua New Guinean Rhythms*. The popularity of this music with audiences saw Tetrafide's arrangement of *Papua New Guinean Rhythms* close almost every Tetrafide concert between 1996 and 2012. The garamuts were also used in new works as a sound source, and later they began to appear in repertoire composed for the group including Pye's percussion quartet *panchavadyam* (2003) for four log drums, metal rubbish bin lids and small concert bass drum.

A desire to treat percussion performances as a style of performance event, rather than a western art music influenced concert influenced Tetrafide's programming towards 2000 (Talbot & Robbie, 2012). In 1998 the group performed a show titled Construction, and although this featured existing contemporary percussion ensemble repertoire that was by this time well established in Australia, the inclusion of lighting and stage design marked the beginning of a change in presentation format for the ensemble, which later led to explorations in other theatrical practices including movement and percussive gesture as the core membership of the group changed. By 1999, Tetrafide included Talbot, Robbie, Richter and Genevieve Wilkins. The incorporation of stage movement is most evident in Richter's quartet for roto toms and woodblocks, *Rough Diamond* (2003).

Around the same time, the exploration of a broad spectrum of global musics emerged in Tetrafide's sound, as music influenced by traditional rhythms of Cuba, Spain and Africa featured regularly in the group's programs, alongside soundscapes evocative of the West Australian regional landscape. Some years later, their programming developed to include commissions using acoustic and electronic percussion instruments by Michael Pignéguy, Steven Richter and Joshua Hogan. This range of influences is reflected in the CD recordings *Tetrafide Percussion* (2004) and *Tetralive* (2007).

A close relationship was formed with colleagues in Nova Ensemble and at WAAPA, and each ensemble was buoyed by a constant sharing of ideas and resources at a local level. Nova and Tetrafide shared a studio in Fremantle for nine years, influencing the creation and evolution of new repertoire as they regularly performed in one another's shows and shared instruments and knowledge gained on various study trips. During this time, collaboration with a range of artists who had migrated to WA from overseas increased and this was reflected in Tetrafide's repertoire. An example can be found in the group's first commissioned work, David Pye's marimba quartet *rebana loops* (2003). Pye acknowledges the influence of his study and collaborations with Indonesian artists in the work in its program note:

> *rebana loops* had its inception in various aspects of Indonesian music, particularly in the technical aspects of the Javanese rebana players. The application of these techniques to contemporary instruments of quite a different kind was the source of inspiration for much of *rebana loops* . . . I find the same

> spirit and energy, sheer enjoyment of music and sense of fun [in Tetrafide] that I have so often found amongst my Indonesian musical colleagues.
>
> (Pye, 2003)

Similarly, Wilkins' study of flamenco in Spain influenced the instrumentation of *Columbia Road* (Blevins & Wilkins, 2006) for marimba, vibraphone, cajon and palmas (clapping), and the instrumentation and harmonic material in Paul Tanner's *Cuba* referenced his five months of study in Cuba and Brazil in 1991. Perhaps the most collaborative project was the 2009 project Spice, that brought together the dancers and musicians of the Temple of Fine Arts, chefs of Indian restaurant Annalakshmi and percussionists of Tetrafide, who at this time were Talbot, Louise Devenish, Joshua Hogan and Marcus Perrozzi. Hogan composed and arranged a number of his earlier works for various drums, marimbas and tabla for this project including trio *Bhaji Smugglers* (2008) to feature Vickneswaran Ramakrishnan on tabla. Perhaps the most successful collaborative work was *Meeting Place* (2009), an exploration of rhythm through instruments, vocal percussion and the footwork of Indian kathak dancers and tap dancer Perrozzi.

In 2000, the group became more active in music education with their program Buildabeat, and was later engaged by Musica Viva in Schools. A successful education program became a major part of the group's output in touring projects throughout regional and metropolitan Australia, Singapore, Vietnam, the Philippines and the United Kingdom. Their education interests were furthered in 2007, with the establishment of the Tetrafide Percussion Advanced Performance Program (2007–2014). Later renamed Tetrafide Percussion Young Artists, this program was designed to encourage secondary school students in both performance and composition for percussion. Over nearly two decades, Tetrafide Percussion had a remarkable impact on the local percussion scene through hundreds of performances in schools that provided the catalyst for dozens of young musicians to study percussion.

Western Australia since 2000

The relative isolation of Perth from other major contemporary arts centres and consequent lack of national and international percussion visitors to Perth influenced early Western Australian explorations of contemporary percussion. Percussionists from this part of Australia had more limited contact with their national peers in the 1980s and 1990s and as a result, a local approach developed that remained dominant for nearly 30 years. Although the influence of international contemporary percussion practices remained evident, two themes can be traced in Western Australian percussion music activity during this time. First, the rhythmic use of found objects and junk percussion instruments. Second, deep exploration into the music of neighbouring countries with a particular focus on Indonesian, Papua New Guinean and Indian instruments and musics. All of the professional contemporary percussion ensembles discussed in this chapter, including Nova Ensemble, AC/PVC and Tetrafide Percussion have explored ways to bring together the instruments of neighbouring countries with contemporary percussion techniques to create new work. This practice continued into the twenty-first century, further developed through the use of electronics as shown in a variety of smaller projects such as duo Taal Naan's studio recording *Rhythm Bred* (2009) and Stuart James' work for Balinese gongs and electronic processing *Kinabuhi | Kamatayon* (2017). The turn of the century also saw another more experimental stream of contemporary percussion music emerge with a focus on graphic notation and electroacoustic works through the artists involved in Decibel New Music (see Vickery, Devenish, James, & Hope, 2017) and theatre-percussion group The Sound Collectors.

References

Amirkhanian, C. 1989. "Lou Harrison and Bill Colvig Interviewed by Charles Amirkhanian." https://archive.org/details/HarrisonColvigInterview

Askill, M. 2017. "Composing with Ancient Sound Technology in the Twenty-First Century." *Contemporary Music Review*, 36 (1–2): 86–101. doi:10.1080/07494467.2017.1368175

Blevins, G. & Wilkins, G. 2006. *Columbia Road.* Perth: Self-published.

Cohn, N. 1986. "Ensemble's Rich Percussion Harvest." *The Australian*, n.d. Accessed David Pye's personal collection.

Cohn, N. 1987. "Pitiless Assault by Percussion." *The Australian*, 22 September.

Meyer, J. A. 1999. *Touches of Sweet Harmony: Music in the University of Western Australia 1953–1998*. Nedlands: CIRCME, University of Western Australia.

Partch, H. 1974. *Genesis of a Music*, 2nd ed. New York: Da Capo Press.

Pye, D. 2003. "Rebana Loops: Program Note." Accessed November 14 2017. www.novaensemble.com.au/davidpye/rebana.html

Ryan, R. 2016. "Forging a Singular Musical Path: The Compositional Explorations of Mark Cain." *Sound Scripts*, 5: 33–43.

Vickery, L., L. Devenish, S. James, and C. Hope. 2017. "Expanded Percussion Notation in Recent Works by Cat Hope, Stuart James and Lindsay Vickery." *Contemporary Music Review*, 36 (1–2): 15–35. https://doi.org/10.1080/07494467.2017.1371879

Personal communication

Cain, M. 2013. Email Communication with Louise Devenish, 3 December. Fremantle, Western Australia.

Maydwell, T. 2012. Email Communication with Louise Devenish, 3 November. Perth, Western Australia.

Pye, D. 2012. Live Interview with Louise Devenish, 24 September. Fremantle, Western Australia.

Sarcich, P. 2012. Skype Interview with Louise Devenish, 11 October. London, United Kingdom.

Talbot, N., and I. Robbie. 2012. Live Interview with Louise Devenish, 10 October. Perth, Western Australia.

Tanner, P. 2012. Live Interview with Louise Devenish, 27 September. Perth, Western Australia.

6 Percussion in Brisbane, Canberra and Hobart

The cities of Brisbane, Canberra and Hobart have played different roles in the development of contemporary percussion in Australia. Prior to 2000, a number of similarities existed in the musical landscape of the three cities, largely resulting from their small populations. This included a focus of activity on the state symphony orchestras and their performers. In Brisbane, Canberra and Hobart in the twentieth century, the majority of percussion activity was in the symphony orchestras, which served as a professional training ground for numerous percussionists following completion of their tertiary studies. However, a high turnover of orchestral percussionists occurred in these cities in the 1970s and 1980s, and as Australian conservatoria tended to employ percussionists who held orchestral positions at this time, this subsequently impacted on the wider percussion scene (Dovey, 2014; O'Kelly, 2013). Opportunities to become engaged in contemporary percussion practices in Australia were often linked to the availability of professional training at tertiary institutions and their instrument collections. It was not until the 1990s that symphony orchestras and institutions in Brisbane, Canberra and Hobart employed percussionists who lived in these respective cities for extended periods, enabling establishment of other contemporary percussion activity. The previous lack of long-term residents resulted in considerably less frequent contemporary percussion activity outside of the symphony orchestras than in other Australian cities. Consequently, any activity that did take place occurred at a grassroots level and largely served as a springboard for future activity by its instigators in larger cities. Although much of this activity was locally restricted, it is documented here in order to present a complete picture of the cultural climate that contributed to Australian contemporary percussion developments in the twentieth century. The catalyst for change came in 1987 in Hobart and Brisbane with the arrival of immigrant percussionists who engaged in percussion ensemble activity at conservatoria in each city. Ten years later, in 1997, a similar change took place in Canberra. This chapter first discusses activity in Hobart and Canberra in the context of the wider Australian percussion scene in the 1990s, followed by in-depth examination of activity in Brisbane that emerged during the same period.

Hobart

Although Hobart is in relatively close proximity to the Australian cities of Melbourne and Sydney, comparatively little contemporary percussion activity originated in this city at the same time that the art form was emerging in other centres. This can be partially attributed to the tertiary training available in Tasmania in the 1980s. The Conservatorium of Music at the University of Tasmania does not house any full-time percussion faculty, and when percussion tuition was made available in the 1980s a series of sessional percussion tutors were employed until 1987, preventing prior establishment of a long-term program (Wain, 2013). In the early 1980s, some contemporary music activity emerged in Hobart, with the arrival of Tom O'Kelly (b. 1961). O'Kelly studied at the Victorian College of the Arts (VCA) in Melbourne, first with John Seal at the VCA Technical School (now the VCA Secondary School), and then as an undergraduate with Barry Quinn, where he was exposed to some contemporary music in addition to his orchestral studies. O'Kelly, however, did not complete his tertiary studies at the VCA, as he accepted a principal percussion position with the Tasmanian Symphony Orchestra (TSO) at the end of his second year of study in 1981. Prior to his move to Hobart, O'Kelly had attended Sylvio Gualda's 1981 concerts in Melbourne discussed in Chapter Two, which O'Kelly cites as inspiring and influential (O'Kelly, 2013). Thus, in addition to commencing with the TSO and teaching a small number of students at the conservatorium, O'Kelly engaged in contemporary chamber music performances with fellow TSO percussionist Malcolm Leek and visitors from Melbourne with whom he had worked as a student, including Australian Contemporary Music Ensemble cellist Christian Wojtowicz and Keith Humble. Although only two or three such performances took place each year, the programs were often ambitious, large in scale and drew a number of performers from around Hobart into contemporary music. Programs requiring numerous percussionists such as George Antheil's *Ballet Mécanique* (1923–24) provided contemporary performance experience for O'Kelly's students. However, O'Kelly left Hobart to pursue study with Gualda in Versailles in 1983, halting these projects for a time.

O'Kelly returned to Tasmania and resumed his positions with the TSO and conservatorium in 1986, where he was able to introduce the ideas and techniques he had experienced in France with a new groups of students. In 1987, Robert Cossom (b. 1967) commenced formal ongoing percussion lessons at the Tasmanian Conservatorium with O'Kelly. Cossom is the only graduate of the conservatorium program to embark on a full-time performance career, and reflects that the training available for young percussionists was limited. As recently as the late 1980s, training was only available via lessons from TSO percussionists or 'on-the-job' with pipe bands or in the various ensembles of the Tasmanian Youth Orchestra (Cossom, 2014). Echoing the sentiments of Adelaide-based percussionists as shown in Chapter Four, Cossom reflects that the small scene in Tasmania and lack of a formal study program including a percussion ensemble was at times advantageous. As there were few guidelines or restrictions surrounding percussion study at the Conservatorium, students were able to experiment with relative freedom.

For Cossom and his peers these experiments included a percussion ensemble; and their student performances in the late 1980s were the first contemporary percussion ensemble activities in Hobart. However, before the end of the 1980s, both O'Kelly and Cossom left Hobart to pursue other professional opportunities with the Orchestra Ensemble Kanazawa in Japan and the Melbourne Symphony Orchestra, respectively. O'Kelly was succeeded in Tasmania by New Zealander Gary Wain in 1987. Wain has held the TSO principal percussionist position and taught tertiary percussion students since this time and his ongoing residency in Hobart enabled him to establish more long-term contemporary percussion activity in the 1990s, including some professional percussion ensemble output with his ensemble Drumwheel.

Drumwheel (1991–1996)

Together with Tasmanian Symphony Orchestra timpanist Duncan Catanach, Stephen Marskell and Johanna Wadsley, Wain established Drumwheel under the auspices of the Friends of Contemporary Music in 1991. Wain states that 'Drumwheel's aim was to play through the standard ensemble rep [sic] of the day and whatever else took our fancy. I also saw it as a way of improving our ensemble playing skills which would be useful when playing as a section in the TSO' (Wain, 2013). By this time the reputations of professional ensembles in other cities such as Synergy Percussion, Nova Ensemble and Adelaide Percussions were well-known around the country and their programming models were mirrored in Drumwheel's concerts. For example, in addition to composing repertoire for the ensemble, Wain programmed Australian works by Sydney-based percussionist-composers Timothy Constable and Michael Askill, as well as international repertoire by Carlos Chavez and John Cage. However, Drumwheel was active for only five years between 1991 and 1996, and performed relatively infrequently. Since the ensemble disbanded, no professional contemporary percussion ensembles have emerged in Hobart in its place, and limited contemporary music took place in Hobart until the early twenty-first century. In 2008, the Museum of Old and New Art: Festival of Music and Art (MONA FOMA) was established, heralding a change in the perception and availability of contemporary performance in Hobart. Although the majority of contemporary music performance presented at MONA FOMA is from cities other than Hobart, the success of this festival highlights the appetite for contemporary performance in Hobart in the twenty-first century. Australian percussionists from Sydney and Melbourne regularly feature at the festival as soloists, percussion ensemble and chamber ensemble performers such as including Speak Percussion, Claire Edwardes and Bree van Reyk.

Canberra

After ten years in Western Australia, in 1997 Gary France moved to Canberra to establish a percussion course at the Australian National University School of Music (hereafter ANU). Previously, tertiary training in Canberra from percussion specialists largely took place when percussionists from neighbouring cities visited, as described by Rebecca Lagos in Chapter Three. France's appointment enabled a dedicated percussion department and student ensemble DRUMatiX to be established, and over the next two decades a range of percussion activity within the university took place, including significant gatherings featuring North American guest percussionists. These included the Australian National Percussion Symposium (2003) and the International Marimba Festival (2001, 2008), which drew large numbers of percussionists and served to link percussion activity in Canberra with that in other cities, and to develop relationships in the wider Australian percussion community in order to examine and reflect on national activity at the time. Although a number of percussionists graduated from France's program at ANU between 1997 and 2015, few remained in Canberra for the long-term, preventing the establishment of ongoing professional percussion ensembles until 2013. At this time, Ensemble Metatone (2013–2016) was established by Charles Martin as part of his research at ANU. Ensemble Metatone was part of a twenty-first-century wave of experimental percussion activity engaging with technology, and the ensemble primarily used an iPad rather than traditional percussion instruments in performance, performing improvisations using iPad applications designed by Martin (Martin, 2017) prior to the departure from Canberra of two of its members.

Brisbane

Like Hobart and Canberra, percussion activity in Brisbane followed a path unique from their neighbours on the East coast of Australia in the twentieth century. Although tertiary training with a focus on orchestral performance was available to percussionists in Brisbane as early as 1970, a high turnover of trained performers in the 1970s and 1980s impacted on the way percussion activity developed in Brisbane during this time, as was the case in Hobart. The following paragraphs outline the approach to percussion training in Brisbane prior to 2000 and the impact of this on local practices. This is followed by an examination of professional activity post-2000, with particular focus on the ensembles Isorhythmos and Clocked Out duo.

In 1970, The Queensland Conservatorium at Griffith University engaged The Queensland Orchestra (now The Queensland Symphony Orchestra, hereafter QSO) principal percussionist Ken Clark to teach percussion as a single study in circa 1970. The year that Clark began teaching at the conservatorium is unknown, as Clark passed away shortly before this research commenced and neither personal or institutional records can confirm the exact date. Clark taught percussion on a sessional basis at the conservatorium until circa 1978, and in 1980 percussion was introduced to the conservatorium's Bachelor degree program with the appointment of Richard Mills. Mills held the position of Lecturer in Percussion for a short time until his departure in 1981, and a number of percussionists, including Graeme Leak, Ross Dovey, Gary France and Gary Wain, followed. In 1987, Alan Cumberland was invited to take up the position of Senior Lecturer in Timpani and Percussion. Cumberland held this position for ten years, providing some stability to percussion education in Brisbane. Prior to this appointment, Cumberland was the principal timpanist of the London Philharmonic Orchestra for two decades and Professor of Timpani at the Royal College of Music between 1980 and 1986. Thus, tertiary percussion education at the Queensland Conservatorium led by Cumberland was firmly based in the English orchestral traditions towards which many Australian music institutions had aspired in the mid-twentieth century. This remained the case until 1998, when Cumberland left Queensland in 1998 to take up a position at the Hong Kong Academy for Performing Arts. Although this represented a sound basis for percussion study, it did not align with the changing Australian contemporary music landscape as there were no contemporary percussion ensemble components to the course. This conservative approach separated percussion activities in Brisbane from those in neighbouring cities at the time, which were decisively including contemporary practices alongside existing orchestral work both within institutions and in the professional sphere.

Following Cumberland's departure in 1998, percussion tuition returned to a single study basis available from sessional staff, and the Head of Percussion position remained vacant for five years (Roennfeldt, 2011: 159). Notwithstanding, the turn of the twenty-first century heralded change in percussion activity in Brisbane with the establishment of Rhythmscape Publishing, Isorhythmos percussion ensemble, and the arrival of Clocked Out duo.

In 2000, Gordon Hughes founded Rhythmscape Publishing, an organisation dedicated to promoting Australian solo and ensemble percussion repertoire. Rhythmscape represents 29 Australian composers from Brisbane, Perth, Sydney, Canberra, Melbourne and Tasmania publishing percussion music by Elena Kats-Chernin, David Pye, Robert Cossom, John Arcaro, Daryl Pratt, Mary Broughton and Kerryn Joyce among others, and publishes new work annually for Robert Cossom's MSO Snare Drum Award. In the same year, Isorhythmos percussion ensemble was established by David Montgomery, who directed the ensemble between 2002 and 2011. Isorhythmos was designed to engage percussionists working with the Queensland Symphony Orchestra in ensemble activity. Operating on a project basis, Isorhythmos drew on a large pool of percussionists from the local scene, including Leah Scholes, Blair Fairbairn, John Douglas, David Kemp, John Parker, Tanja Hafenstein and Andrew Knox. Their programming reflected the established Australian percussion ensemble format that juxtaposed seminal international repertoire with Australian repertoire made popular by Synergy Percussion, music influenced by global percussion traditions and commissions for the ensemble, including works composed by Montgomery and other members of the ensemble. Like their predecessors the Australian Percussion Ensemble, Adelaide Percussions, Synergy Percussion and Nova Ensemble, they regularly performed with guests, programming mixed chamber ensemble repertoire as well as percussion ensemble repertoire. Isorhythmos engaged with the local scene, and collaborated with a number of Queensland-based composers on new works, including Gerard Brophy's marimba solo *Trance Ripples* (2002) and a number of works by Alan Lawrence. Notably, in 2005 Isorhythmos presented an entire program of music by Queensland composers Lawrence, Michael Dixon, Peter Rankine,

Erik Griswold and John Gilfedder. With the exception of Griswold's sextet *In the Dream* (2004), all of the works on this program were commissioned for the event. However, the majority of these works have not been performed outside Queensland, or by other ensembles.

Vanessa Tomlinson and Clocked Out (1999–present)

Tomlinson's early contributions to Australian percussion in 1990s Adelaide were discussed in Chapter Four, and since that time Tomlinson has premièred over 100 new works. Tomlinson had returned to Australia in 1999 following six years of study with Steven Schick at the University of California San Diego, the same institution that Jean-Charles François taught percussion following his departure from Melbourne in 1971. Tomlinson was initially based in Melbourne and for three years was extremely active in contemporary music performing as a soloist, performing with percussion quartet Playing Fields and Clocked Out duo with pianist-composer Erik Griswold, and curating three concert series: Virtuosic Visions, Free Art Melbourne Experiments (colloquially referred to as FAME) and Dada Cabaret. These activities will be briefly discussed here, as they formed the basis for future activity in Brisbane.

Playing Fields was an experimental percussion quartet comprising Tomlinson, Graeme Leak, Peter Humble and David Hewitt. Although active only in 2001 and 2003, Playing Fields' output was somewhat separated from other percussion ensemble activity during the same period due to the extent of influences from the theatre arts, improvisation and group composition appeared in their practice. Repertoire by Feldman, Cage, Ferneyhough and Globokar featured in their programs, however their most notable repertoire was composed by members of the ensemble. This included group compositions that incorporated movement, spoken text and theatricality, and works that were designed to explore the acoustic sound within the physical performance space. In one review, O'Connell remarked that 'what one saw was just as interesting as what one heard' (O'Connell, 2002). For example, Peter Humble's *Father and Son* (2002) featuring a tape loop of live sound reminiscent of lowing cattle was performed by Humble while his father (also named Peter Humble) delivered an auctioneers spiel, treating the audience as a crowd bidding on cattle. Playing Fields also collaborated with a range of performing and visual artists, including sculptor Rosemary Joy, who explored some of her earliest sculptural percussion work with the ensemble. Joy describes her percussive sculptures as a medium that physicalises sound, and states that their performance creates a time-structured experience of the sculpture informed by its appearance and the space in which it is performed (Joy, 2015: 13). An example of an early percussive sculpture by Joy for Playing Fields is *Field Kits* (2003), a work that includes four miniature travelling percussion kits personalised for each member of the ensemble.

Virtuosic Visions I and II were experimental site-specific recital series curated by Tomlinson and Griswold in 2001 and 2002. Held in various spaces at the Melbourne Museum, each recital was designed to interact with the architecture and exhibitions of the time, and featured a range of Melbourne-based artists working in improvisation, jazz, punk and experimental musics including performer-composers Natasha Anderson, Robin Fox, Erkki Veltheim, Tomlinson, Griswold and others. Within this series Tomlinson gave solo recitals of repertoire by collaborators from Melbourne and San Diego such as Warren Burt's *Beat Generation the California Coastal Ranges* (1999) for vibraphone and sine tones and her own compositions *Dear Judy* (2001) for solo balloon and *No* (2002). Following on from Rhythm Method's concerts in the 1990s, Virtuosic Visions I and II demonstrate that unconventional performance in settings beyond the concert hall has been a key part of Tomlinson's practice for some time, and this continued to develop in scope following her relocation to Brisbane in 2003.

In 2003 Vanessa Tomlinson was appointed Head of Percussion at the Queensland Conservatorium Griffith University (QCGU). Tomlinson's arrival commenced a period of significant change in direction for percussion activity in Brisbane, primarily through her activities at QCGU. This included establishing Ba Da Boom, a student percussion ensemble that focusses equally on improvisation, new works and performance of existing international and Australian repertoire. Tomlinson reflects that the ensemble's name is in homage to Vinko Globokar:

> Globokar wrote an article called *Anti-badabum* (Globokar, 1992), where he was really against this idea of percussion being purely driven by speed and technique and not by awareness of sound. When I started the ensemble, I decided that we were beyond that 'anti-ness' and I could call it Ba Da Boom in honour of him.
>
> (Tomlinson, 2014)

Although Clocked Out originally commenced activity in Melbourne in 1999, the ensemble has since had significant impact on Australian percussion music in Brisbane since 2003, when they became ensemble-in-residence at QCGU. Clocked Out's work has combined influences from jazz and world music traditions, the experimental Dada and Fluxus schools, sound installation, acoustic ecology and site-specific work using prepared piano, percussion and toy instruments. Improvisation and collaboration are central to the ensemble's approach and the arrival of this duo on the Brisbane percussion scene marked a distinct change in direction from previous percussion activity. The dual combination of playful and serious exploration is a hallmark of Clocked Out projects. This duality is a deliberate approach, described by Griswold thus:

> We want to set up an environment where it will seem natural that a complex, sophisticated piece can come up next to a pop song. There's a certain amount of nostalgia there, going back to a time before the split between high art and low art became such a dominant part of culture.
>
> (Griswold in Smith, 2003)

Goodbye Elvis (2008), a program of works by Xenakis, Takemisu, Scelsi, Queen and The Sex Pistols interspersed with 1970s television commercials, is one example of this approach. Intrinsic to their exploration of sound is a theatrical sense of play that draws audiences into contemporary music through site-specific projects, and collaborations with community and professional musicians. In stating that 'Clocked Out exists both as an investigation of percussion and also investigation of sound in place' (Tomlinson, 2014), Tomlinson highlights two important themes in her work in the context of Australian contemporary music. In Clocked Out, one method of interacting with their Australian context is the transformation of sounds and concepts from everyday life through performance. This is achieved via the use of urban materials but also in site-specific projects that engage with urban and natural environments such as the 'Sounding' series. This series of site-specific, community-based works have exploratory investigations of sound in place at their centre: *Sounding Wivenhoe* (2007), *Sounding Nudgee* (2008), *Sounding the Condamine* (2009) and *Sounding Noosa* (2017). *Sounding the Condamine* is particularly remarkable, as it celebrated the sound of a uniquely Australian metal bell and invited listeners to bring their own historic Condamine bells to sound during the event. An adaptation of British bells worn by cattle to locate them as they roved the countryside, the Condamine bell was developed by Queensland drovers in the 1800s to produce sound that carried over large distances in regional Australia (Cooney, Knie & Knie, 2008). Until *Sounding the Condamine* these bells had not previously entered Australia's art music landscape. Tomlinson reflects on the investigation of sound in Australia as a significant part of her practice:

> I still feel like I am the most privileged human in the world to be able to use Australia as my sound laboratory. I can go out investigating so many different layers of sound and different possibilities of listening with unbelievable ease, just living where I do now, but also living in Melbourne or living in Adelaide. I'm particularly interested in place and landscape and soundscape.
>
> (Tomlinson, 2014)

The work of Clocked Out in the first decade of the twenty-first century sparked a trend in experimental music practices featuring diverse approaches to percussion that have continued to develop in recent years. Smith states that 'the remarkable thing about Clocked Out Duo's performances is not so much their wild eclecticism, but the way they maintain a strange cohesiveness and integrity' (Smith, 2003), and although the influence of Clocked Out has seen a shift in percussive music-making practices in Brisbane against what has been described by Hooper as 'Brisbane's relentless normative push' (Hooper, 2012). The evolution of Queensland-based percussion practices is further evidenced in Early Warning System, a percussion ensemble co-founded by Tomlinson and Michael Askill in 2011. With the bold catchcry of 'changing the sonic landscape of percussion in Australia', the ensemble takes its name from Marshall McLuhan: 'art at its most significant is a distant Early Warning System that can always be relied on to tell the old culture what is beginning to happen to it' (Early Warning System, 2017).

Although initially less active than other Australian centres prior to 2000, percussion activity in Brisbane has substantially increased since the turn of the century and is now an established hub. This is demonstrated through three major events have taken place in Brisbane at QCGU since 2010, that have brought the majority of the Australian percussion community together: the Australian Percussion Gathering (2010, 2016) and Transplanted Roots Percussion Research Symposium (2017). These events coupled with recent ensemble activity suggest the twenty-first century will reveal a rich body of Brisbane-based creative work worthy of future research.

References

Cooney, D., E. Knie, and P. E. Knie. 2008. *Bells of the Australian Bush*. Toowoomba: Jinglestix.

Early Warning System. 2017. "About." Accessed 7 December 2017. www.earlywarningsystem.org.au/about

Globokar, V. 1992. "Anti-Badabum." *Percussive Notes*, 31 (1): 77–82.

Hooper, G. 2012. "Bracing New Music and a New Percussion Ensemble." *RealTime*, 107: 42.

Martin, C. P. 2017. "Percussionist-Centred Design for Touchscreen Digital Musical Instruments." *Contemporary Music Review*, 36 (1–2): 64–85. https://doi.org/10.1080/07494467.2017.1370794

O'Connell, D. 2002. "Father and Son Make Cattle Music." *The Age*, 17 September.

Roennfeldt, P. 2011. "Beyond the Ivory Tower – Higher Education Institutions as Cultural Resource: Case Study of the Queensland Conservatorium of Music." *History Compass*, 9: 231–245.

Smith, R. 1990. "Skilful Sounds Tailored to Surroundings." *The Advertiser*, 25 August.

Smith, R. 2003. "How the Balloon Taught the Piano to Play." *RealTime*, 55: 28.

Personal communication

Cossom, R. 2014. Telephone Interview with Louise Devenish, 1 March. Melbourne, Victoria.

Dovey, R. 2014. Email Communication with Louise Devenish. Brisbane, Queensland.

O'Kelly, T. 2013. Live Interview with Louise Devenish, 22 November. Brisbane, Queensland.

Tomlinson, V. 2014. Telephone Interview with Louise Devenish, 5 July. Brisbane: Queensland.

Wain, G. 2013. Email Communication with Louise Devenish, 17 April. Hobart, Tasmania.

7 The question of identity

Looking back, looking out, looking in, looking forward

When considered collectively, the previous chapters demonstrate that Australian contemporary percussion has always been about more than just Australian music; it has consciously connected with global practices due to its time and place in music history and to Australia's cultural context. Thus, documenting the history of contemporary percussion practices in Australia between 1960 and 2010 brought a number of questions about Australian percussion identity to light. Are there any defining characteristics to be found in the body of Australian percussion music? If so, are they characteristics unique to Australian percussion? Have Australian practices aligned with any international practices? How have advances in technology affected the development of global schools of thought? How do Australian percussionists view their practice in local and global contexts? These questions give rise to others: are questions of identity relevant to the creation of new contemporary percussion work?

This chapter seeks to introduce questions of identity such as these at a turning point in the evolution of contemporary percussion, when scholarly research in the field is diversifying. It does not attempt to analyse the influences presented here in significant detail. Rather, it offers preliminary responses based on findings from the previous chapters and invites debate and further research.

Given that the dates of this study align with a period in Australian musical history that saw a conscious break from the previous inextricable links with English practices, that questions of identity surfaced was inevitable. However, during the course of over 50 interviews with Australian percussionists and their collaborators, it transpired that questions surrounding identity are not often explored within the Australian percussion community. When such questions are posited, the evolving nature of contemporary percussion renders succinct answers difficult. Questions of percussion identity are inherently linked to the question of what percussion is – and what it could be. Ughetti states that 'the physicality of how we approach our instrument is in flux, the timbres that we make are in flux, the performance techniques are constantly changing and rapidly evolving' (Ughetti, 2014a). Joy suggests that this state of flux is a result of percussionists' flexibility in approach beyond instruments and technique, observing that this flexibility 'extends to a willingness to engage with new ways of presenting work and new ways of thinking about percussion' (Joy, 2015). Willingness and creative flexibility has long been a prerequisite for successful performance of percussion repertoire, and is often the most vital skill required from performers. Schick outlines this using Stockhausen's *Nr. 9 Zyklus* (1959) as an example: 'The piece, in essence, doesn't demand anything more than an open form of reference and an ability to solve the problem' (Schick, 2015). As a result, astounding diversity is present within the field of contemporary percussion. Many varied approaches to performance, repertoire and education thrive alongside one another in all international centres where percussion practices exist, making any concise articulation of the art form challenging. Gottschalk states that the wider genre of experimental music is similarly difficult to articulate:

> It is a term that refuses to be defined. Experimental music resists, stretches, and breaks boundaries. Defining it is like mapping a beach. The water line changes with the tide.
>
> (Gottschalk, 2016)

Despite this (or perhaps because of this), questions of identity are being asked with increasing regularity in the wider contemporary percussion community as the art form enters a new phase in its history. This discussion includes examination of percussion repertoire, which has 'evolved to the point where a historical perspective regarding the defining characteristics of various works [has] begun to cohere' (Palter in Sallak, 2014: 189), and examination of percussionists and performance practices (see Kernan, 2014; Schick, 2014). How practitioners reflect on and engage with history through the creation and new work and performance of existing work is a crucial part of moving forward, whether or not this includes conscious consideration of questions of identity. As Fox states, 'addressing questions and examining roles is central to making work in an experimental way' (Fox in Carnegie, 2014). Recent publications have sought to contribute to research on percussion inclusive of percussion history, artistic research practices and the emerging field of intercultural music making (Hartenberger, 2016; Schick, 2006a; Lewis & Aguilar, 2014; Devenish, 2017). In 2015, the biennial Transplanted Roots Percussion Research Symposium was established, in part to ask questions about the past, present and future of percussion. Whether or not the questions posed by this research can be answered is yet to emerge, but it is clear that the rapid rate of change set in the twentieth century is not slowing.

On a global scale, we can see that the definition of a western percussion instrument is constantly changing. For example, part of what makes an Indonesian gamelan ensemble sound Indonesian, or a Korean drumming group sound Korean, is the type of instruments and instrumental combinations that can be used in these ensembles. Specific instrumentation is crucial in establishing identity for many global percussion ensembles, and for western chamber ensembles developed in centuries past, such as the string quartet. Contrastingly, many western contemporary percussionists across numerous continents use a wide variety of instruments drawn from numerous sources around the world, and every contemporary percussion ensemble uses a unique combination of instruments. Further, percussion instrumentation is not restricted to musical instruments, but extends to any object from which sound can be drawn. The instrumentation of four flowerpots in Frederic Rzewski's *To the Earth* (1985) has become a famous example of this. Unlike other western art music instruments, the development of percussion has not been primarily focussed on perfecting the design of a small number of instruments – as occurred with string instruments, the design of which has not changed substantially for hundreds of years. As Cleworth states, 'part of the way percussion has developed is, pure and simple, just about sound sources' (Cleworth, 2013). In the search for diverse sound sources and new ways to make music, percussionists have become adept at bringing together of a range of sound sources, and this has had an impact on the development of technique and performance practices. Contemporary percussionists have sought to develop an approach to music making that enables the adaptation to a range of instruments and objects, rather than refining technique for a single instrument. Further, many have looked beyond music practices for inspiration, and the incorporation of aspects from the visual and theatre arts, sound installation practices, instrument building and technology is now common. There are now numerous practitioners around the world whose experimental percussion work lies at the nexus of multiple art forms. Identifying as percussion music but without using recognisable percussion instruments or techniques, this has been termed post-instrumental or post-percussive practice by Håkon Stene (Stene, 2015: 2). In other cases, repertoire has entered the percussion canon by virtue of the lack of boundaries around what constitutes percussion music, such as Mark Applebaum's *Aphasia* (2010), which was originally intended for singer and tape. This work has become a piece of percussion repertoire by virtue of the fact that percussionists perform it, not because of its instrumentation or performance techniques, a notion deliberately explored by Rosemary Joy's 2011 work *The Appearance of Sound* (Joy, 2015). This global expansion of what constitutes a percussion instrument means that any discussion of identity – Australian or otherwise – must be of a wider scope than instruments and technique to include consideration of local history, influences on repertoire, performance practices and musical philosophy.

Looking back

When discussing the place of Australian contemporary percussion in a global context, the constant state of evolution coupled with the relatively short Australian history presented a hurdle for some interviewees in considering identity. Some suggested that a history dating from the 1960s could not provide enough data to reveal an Australian percussion sound or identity, particularly given the aforementioned rate of change. However, as exploration is the epicentre of percussion practice, it can reasonably be assumed that this state of perpetual evolution is indefinite, thus the short history of percussion in Australia is perhaps not reason enough to forestall the question. It could be further argued that engaging in critical reflection at a time when many Australian percussion pioneers are able to contribute first-hand experiences to the narrative is precisely when documentation and discussion of twentieth-century practices should take place. The importance of primary contributions was highlighted in personal communication with American percussionist Allen Otte about John Cage's percussion solo *27'10.554" for a Percussionist*. Composed in 1956, *27'10.554" for a Percussionist* was not premièred until 1962. Although it was published in 1960, there is a gap of six years between composition and performance, and there is no indication for whom this work may have been composed in the score or archival material surrounding the première. Otte worked extensively with Cage, and has since engaged in research on Cage's work via archives and catalogues. When questioned about the work's origins, Otte's reply began 'It would be a good subject for an article: All the questions I wished I had asked [John] Cage and [Lou] Harrison and Herbert Brün'. He concluded with the suggestion: 'keep asking questions . . . you'll regret what you didn't ask' (Otte, 2013).

Another reason interviewees cited for their reticence to discuss identity was the intangible nature of music. When discussing identity in terms of sound, the temporality of musical sound reduces the effectiveness of language. A number of practitioners suggested that an Australian contemporary percussion sound was not something that could be codified easily:

> I'd be hard pressed to say that there is such a thing as an Australian percussion sound, I'm not sure that actually exists, . . . I think it's difficult to pin down or define an Australian percussion sound, . . . However, I think ensembles and players can develop a particular sound . . . words are very difficult when it comes to describing sound.
>
> (Cleworth, 2013)

> I feel like, to give a simple response is to deny the truth. I've heard it said that sensitivity and an ability to listen and absorb other cultures is part of it . . . but I've only heard that from people in other countries. I've had very few discussions of any part of Australian sound with Australians. I have some sense that we almost talk ourselves out of it somehow.
>
> (Constable, 2014)

> Whether we could recognise an Australian sound, I don't know. I am not confident that I would be able to tell Australian percussion music in a blind test.
>
> (Scholes, 2014)

> I don't think Australian percussion music can be codified easily. I see it very much as a multitude of influences . . . there is not an overarching or defining language used by Australian composers and I think that's very indicative of the cosmopolitan country that Australia is.
>
> (Ughetti, 2014a)

> I don't feel we have an Australian sound, I think the notes speak for themselves. But [in the 1990s] I think we did. For a tiny slice of history we had a

> sound which was coined by Martin Wesley-Smith, Carl Vine, Nigel Westlake: the Flederman guys. They all wrote for Synergy, and they defined a sound.
> (France, 2013)

> I'm not sure whether it's a good thing or not to have a national percussion sound, but I can't see that there is one in Australia . . . we've been so open to outside influences for so long that we're this crucible of sounds and ideas from other places and players have made their own mixture . . . what you do see is an attitude of openness and interest and such a variety of approaches.
> (White, 2014)

The preceding comments also reflect elements of a wider discussion of identity in Australian music that refers to the wide scope of influences on Australian composers since the 1960s. The validity of this discussion is challenged by some composers and musicologists who suggest that 'music is music' and the search for an Australian musical identity should not be a focal point in the creative process (Logan, 2007: 35–50). It cannot be disputed that superficial attempts to create an identifiable Australian sound are artistically questionable. While there is no evidence to suggest a conscious effort to create Australian-sounding percussion work, there is evidence to suggest identity remains important to practitioners. Examination of Australian percussion activity since the 1970s reveals the active championing of Australian repertoire by a significant proportion of percussionists around the country, and it remains a key focus for a range of professional practitioners, including Synergy Percussion, Speak Percussion, Vanessa Tomlinson, Claire Edwardes, Louise Devenish and many others. This stems primarily from a desire to connect with community through collaboration rather than a desire to develop Australian-sounding music. Ughetti explains his engagement with Australian artists thus: 'the idea of creating Australian and even local music is very important to me because I want to be feeding and I want to be fed by the culture I live in' (Ughetti, 2014a).

Although an Australian percussion sound may not be either possible or desirable to define, consideration of Australian percussion repertoire and artistic practices in the twentieth century reveals common themes. These themes are drawn from a range of both external and internal influences that speaks to the wider cultural context and time in which Australian contemporary percussion emerged. In centuries past, geographical boundaries affected musical and cultural development. Travel between countries was arduous, time-consuming and financially challenging, and thus less frequent than it is amongst musicians in the twenty-first century. Irregular exposure to work in other centres of activity enabled practitioners in one country to develop practices with recognisable sonic differences from those in another. Current practitioners of instruments that reached their peak development in centuries past, such as the oboe or piano, are able to identify sound associated with particular schools of thought, which are often codified geographically.

Contemporary percussion emerged at a time in Australian history when an increasing interest in multiculturalism and the concept of globalisation was at the forefront of the national consciousness. By the 1990s, when contemporary percussion practices were becoming more established, the advent of the internet improved the speed and frequency of communication between Australian and international percussionists. Additionally, the relatively low cost of air travel increased the ease with which artists from around the world could share knowledge. Access to a diverse range of global musics was welcomed by Australian percussionists, and the open embracing of influences from a range of external sources accelerated development of Australian percussion practices beyond a rate possible even 20 years earlier.

This is reflected in the body of Australian percussion repertoire created prior to 2000 in particular. Here, a number of external and internal influences can be observed, rising to prominence at various points in time as groups of individuals sought to establish unique identities. For example, French avant-garde and experimental music influences appear in repertoire composed for the Australian Percussion Ensemble in the 1970s and in the percussion music premiered by the ACME New Music Co. in the 1990s. In 1990s Sydney and Perth, influences from Asian countries of the Pacific Rim were significant in works programmed by Synergy Percussion, Taikoz, Nova Ensemble and Tetrafide Percussion. It is pertinent to note that these influences are not mutually exclusive, nor are they necessarily unique to Australian percussion music, as they can be found in the repertoire of international ensembles active during these periods. The impact of external and internal influences on Australian percussion practices is discussed next: first, the external – international contemporary percussion in Europe and North America, prior to the impact of Asian and African traditional instruments and musics on Australian percussion repertoire; second, the internal – geography and a sense of place, urban influences and approach.

Looking out

Europe and North America

Chapters One and Two demonstrate that the isolation of Australian cities from contemporary music hubs in Europe and North America was keenly felt by Australian percussionists in the 1970s, with discussion of the catalytic effect of tours by established international artists. Given the reflections in earlier chapters on Les Percussions de Strasbourg tours, it is unsurprising to find that French percussion practices were a primary influence on early Australian activity. This was reflected in names selected by Australian ensembles established in the 1970s, such as The Sydney Percussions and Adelaide Percussions, which were consciously modelled on Les Percussions de Strasbourg despite the rather awkward English it produced. An ongoing connection was established through Michael Askill's study with members of the ensemble in Strasbourg in the 1970s before his return to Australia. The programming of works championed by Les Percussions de Strasbourg by Australian ensembles is another indicator of their influence, although given the dearth of international contemporary percussion repertoire of the time and their status as the world's first professional contemporary percussion ensemble, this is perhaps to be expected.

Keith Humble and Jean-Charles François' work in Melbourne also contributed to the French influence on Australian percussion over a period of a few years while employed at Latrobe University. Their experimental ensemble KIVA and other acoustic and electronic music activities including composition, improvisations and performances of notated repertoire at the time, were informed by their study and work in Paris directly before their relocation to Melbourne in the late 1960s. Further, François and Humble engaged in these activities and ongoing discourse with colleagues at Latrobe. When percussionists from around Australia travelled from neighbouring cities to attend Sylvio Gualda's masterclasses and performances in Melbourne in the 1980s, this signalled an ongoing interest in French percussive practices, complemented by trends in overseas study amongst Australian percussionists. Although few Australian percussionists pursued further international study in the 1970s and 1980s, those who did invariably chose to study with the French artists whose technique and ideas they had been exposed to in Australia. These individuals, namely Tom O'Kelly and Greg Sully in Versailles and Peter Cobb's study Strasbourg. O'Kelly reflects on Gualda's visit to Melbourne, and the subsequent impact this had on his career:

> Sylvio came out, I heard him play not only timpani but everything else, I'd never heard anyone play with such a beautiful sound on everything . . . I couldn't believe this guy so I had a lesson with him, and he said 'you can come and study with me'. But then I got a job in Tassie [sic] and I never heard from him. And then out of the blue, I got a card from him, when I was in the orchestra down in Tassie, saying 'come next month' . . . so I resigned from the orchestra and went.
>
> (O'Kelly, 2013)

O'Kelly studied with Gualda at the Conservatorium of Versailles between 1983 and 1985. At Versailles, O'Kelly came into regular contact with major figures in international contemporary music of the time through Gualda, such as Jacques Delécluse, Toru Takemitsu, Iannis Xenakis, Pierre Boulez and Luciano Berio. For O'Kelly, working with Xenakis on his masterwork *Pléiädes* when the composer was in residence at Festival D'Aix En Provence had a major influence on his development as a musician. During this time, O'Kelly became aware of differences in approach between the European percussion scene and the Australian percussion scene, most notably the relationships between cities. In Australia, when a major percussion event occurred, percussionists would travel long distances to be involved, however in France, O'Kelly perceived that as there were so many more percussionists in each city and therefore a local community, travelling long distances to see their colleagues in other centres perform was rare (O'Kelly, 2013). Despite the fact that Strasbourg was only three hours away from the Conservatorium of Versailles, the percussionists in each city did not regularly interact with each other at this time.

By the 1990s, a new generation of percussionists had come of age. These percussionists repeatedly cite the work of two others as influential on their practice: American percussionist Steven Schick and Swiss percussionist Fritz Hauser. Schick's 1990 tour had a similarly widespread impact as Les Percussions de Strasbourg had nearly 20 years earlier, this time in the context of solo performance:

> Steve Schick coming here was a pretty important moment for quite a lot of percussionists, but certainly for me. I would have to say that before that point, I don't think I had ever seen a solo percussionist before. I mean a solo percussionist that wasn't a student doing a recital, or wasn't me, it was the first time I'd seen a model of what I was trying to be. . . . I'd never really come across that kind of percussionist before.
>
> (Tomlinson, 2014)

> Steve Schick has had a major influence in Australia . . . back from the time when a lot of people weren't coming here . . . he's had a major influence just in terms of that open minded way of thinking about percussion.
>
> (Edwardes, 2013)

> [The 1990s] was a really interesting time. I had never seen anything like Steve Schick before.
>
> (Grigg, 2017)

Schick's long-term influence is evidenced in both the number of Australian percussionists who have spent time studying with Schick since the 1990 tour and Schick's subsequent visits to Australia over the next three decades including the 2003 Percussion Research Symposium in Canberra, the 2010 Australian Percussion Gathering in Brisbane, Transplanted Roots 2017 in Brisbane and the 2017 American Triptych in Melbourne.

During the 1990s Fritz Hauser became another significant influence on Australian percussionists, stemming from an array of professional collaborations with different members of Synergy Percussion, Speak Percussion, Aphids, Chamber Made Opera and independent artists such as Rosemary Joy. Ughetti states that Hauser's first visit to Australia 'changed my worldview at the time . . . he has been a long-term collaborator since then' (Ughetti, 2014a). The formation of ongoing relationships between visiting international percussionists and the Australian percussionists whom they engaged was not uncommon and speaks to the importance of long-term connection and community as a means to lessen the impact of geographical isolation, and as a means to establishing identity.

Asia and Africa

In 1948, Henry Cowell predicted that 'future progress in creative music for composers of the western world must inevitably go towards the exploration and integration of elements drawn from more than one of the world's cultures' (Cowell,

1948: 412). Contemporary percussion has roots in ancient drumming traditions around the world, and many percussionists are drawn to non-western musics in pursuit of understanding the lineage of the art form, and to make links with a global percussion tradition. Discussed above is the central role of exploring sound sources in the development of contemporary percussion practices. From this perspective, the introduction of traditional instruments from world cultures could be considered a natural extension of western percussion playing, as the absorption of sound sources removed from their cultural context and explored purely for their sonic capabilities is part of how contemporary percussion has developed.

The influence of Asian traditional instruments and musics has long been present in contemporary percussion music. This was first manifested in the inclusion of various traditional instruments in percussion ensemble compositions, not dissimilarly to the earlier expansion of the instrumentation of an orchestral percussion section. For example, many types of gongs and tam-tams from various Asian countries, such as the Chinese *chao* gong are by now unquestionably standard items in western percussion studios. By the 1990s, influences beyond instrumentation from non-western cultures was common in projects by numerous high profile contemporary percussion ensembles of the time. This was frequently as a result of extensive travel and study by one or more members of any given ensemble. For example, it was during this decade that Amandinda (Hungary) explored Pacific island drumming and NEXUS (Canada) explored African and Indian percussion traditions and composed repertoire influenced by these experiences. Invariably, the interests of performers also influenced a number of composers who wrote for them. The increasing inclusion of study in non-western percussion musics by western percussionists saw North American universities expand their percussion programs to offer study in music from Mexico, Brazil, Africa, Cuba and Indonesia amongst others. The benefits of engaging with traditional percussion musics and the impact of doing so has been discussed by percussionists globally (Williams, 2016); the best practice for doing so has evolved in the twenty-first century, particularly as the field of intercultural music making grows. As early as the 1960s musicologists observed that the percussionist's idiom 'owes as much to non-western traditions as to western classical music' (Schwartz, 1961: 9).

In Australia, the most common non-western traditions found in contemporary percussion practices are from the neighbouring Asian countries of Indonesia, Papua New Guinea, Japan, China and South Korea, a practice foreshadowed by trends in Australian music in the 1960s and 1970s. Before contemporary percussion emerged, Australian musicians of all disciplines were gravitating towards Asia as a source of inspiration (Rogers, 2009: 257). In the first half of the twentieth century, Australian composers primarily mirrored trends in European music. Changes in the Australian musical landscape in the 1960s led to increased recognition of Asia's proximity, and composers began to seek simultaneously to identify less with Europe and more with the region of which they were geographically a part (Rogers, 2009: 247). The significance of this trend resulted in the foundation of the Australian Music Centre Asian Music Studies Programme in 1975. The influence of Asian musical instruments and practices in solo and ensemble percussion work can therefore be considered a continuation of wider investigations commenced in the 1960s. As discussed in previous chapters, groups of individuals in different Australian cities have connected with the music of neighbouring countries at various times since the 1980s. Many acknowledge their initial interest in these musics as a by-product of being percussionists:

> As percussionists, we [Synergy Percussion] were passionate about the roots of drumming and tried to find an approach to the interpretation of contemporary percussion music that would reflect our location within the Asia-Pacific region.
>
> (Askill in Schaefer, 2007)

> [My interest in world music] started because I was a percussionist.
>
> (Reeves, 2015)

> It was a natural extension for a lot of percussionists then to literally introduce the sounds and rhythms and pieces from other cultures.
>
> (Cleworth, 2013)

> One of the bonuses of being a percussionist is being able to discover the music of other cultures and incorporate it into your playing.
>
> (Talbot & Robbie, 2012)

When Australian percussionists began travelling to Asian countries to study traditional drumming and to purchase instruments, the effect on Australian practices was immediate, as these percussionists shared new knowledge with peers and facilitated communication with musicians in Asia. This in turn influenced the creation of new repertoire, particularly percussion ensembles. For example, the filtration of Asian musical influences into Nova Ensemble's practice was led by Ron Reeves. Reeves had engaged in extensive study in Indonesia and India, and continues this study today. Musical influences from around the Pacific Rim began to appear in music composed by Reeves and other WA composers in one of three ways (Pye, 2012): first, via small ensemble works scored for a combination of Asian and western instruments, as in Stuart Davies-Slate's *Nupela Garamut* (1988); second, via the combination of non-western instruments with western compositional techniques, as in Lee Buddle's *Just an Inkling for an Angklung* (1990); and third, via the influence of Asian musical patterns on compositional structure, such as the application of interlocking kecak rhythms to western concert marimba and tuned concert tom-toms in David Pye's *rebana loops* (2003) (see Example 7.1).

The performance of non-western music from neighbouring Asia-Pacific countries also emerged in Australian percussion practices in the 1990s. In the case of Tetrafide Percussion, this became crucial in contextualising the group as an ensemble of the Pacific Rim through the performance of Papua New Guinean rhythms. As discussed in Chapter Four, Robbie's study of musical traditions and rhythms of various parts of Papua New Guinea, the Cook Islands and Fiji in the 1990s saw the

Example 7.1 An excerpt from David Pye's *rebana loops* showing the application of interlocking material on tuned tom-toms and marimba.

Source: Used by arrangement with Rhythmscape Publishing.

Papua New Guinean *garamut* become as integral to Tetrafide's instrumentation as the concert marimba. The work of northern New South Wales-based musician Greg Sheehan is also heavily influenced by the rhythms of non-western cultures. Over a period of four decades, Sheehan has refined a unique approach to rhythmic composition and improvisation known as number diamonds. Using numerical approaches to composition and the rhythms of a range of non-western traditions around the Asia-Pacific as a point of departure, Sheehan has shared this practice with others since the 1970s through his ensemble Utungan Percussion and independently, encouraging the use of number diamonds in new work (O'Neill, 2013). Westlake acknowledges elements of Sheehan's approach to rhythm as an influence on his early percussion music via collaboration in Utungan Percussion and the Magic Puddin' Band in the 1970s (Westlake, 2017). Four decades later, Sheehan's number diamonds were also applied to repertoire composed by Joshua Hogan for Tetrafide Percussion and Taal Naan.

In the 1990s, Australian projects replete with Asian elements began to emerge, including Matsuri (1992), Synergy Percussion's first major work integrating contemporary percussion and Japanese traditions, and Nova Ensemble's Indonesian inspired project Junkelan (1994). As discussed in Chapter Three, the foundation of Taikoz in 1997 marked a new period of engagement with Asian music, specifically the *wadaiko* music of Japan. Taikoz's increased concentration on the integration of Japanese and western percussion practices signalled a change in the way Australian percussionists were engaging with non-western musics. Moving away from performance of traditional music or use of traditional instruments purely as sound sources, Taikoz's approach to creating new music equally informed by contemporary percussion and Japanese *wadaiko* aesthetics places their output in a unique space. Cleworth asks: 'When can taiko just be music?' He posits that any genre of music must be able to express something other than itself, and this belief is a driving factor in his creative process (Cleworth, 2017).

Towards 2000, Chinese influences surfaced in Australian percussion. In 1999 Erik Griswold and Vanessa Tomlinson (Clocked Out duo) received Asialink Arts funding to visit the Sichuan province to study the local folk and opera percussion traditions, the first of many intercultural exchanges over subsequent years. Asialink was established as a joint initiative between the Australian Government's Commission for the Future, the Myer Foundation and the University of Melbourne in 1990, and has supported numerous artistic residencies in Asia since this time. For example, the introduction of Chinese opera percussion into Clocked Out's practice lead to the creation of The Wide Alley (2007). The Wide Alley blends Sichuan opera percussion, Qin Yin style singing and Sichuan street sounds with influences from jazz, free improvisation and contemporary western art music. Griswold writes:

> Sichuan Opera is known for its characteristic gao qiang style (one of five main percussion styles) of using percussion and voice with no other accompanying instruments. To our ears, trained in contemporary western music, gao qiang style sounded quite avant-garde, reminding us of Luciano Berio or Stockhausen.
>
> (Griswold & Tomlinson, 2010)

For Synergy Percussion, collaboration with world music masters from Europe, South America, Africa and India became a major focus in the 1990s, and intercultural music making has remained a pivotal part of the ensemble's practice into the twenty-first century. In 2010, for example, Timothy Constable was the featured composer at the Seoul Drum Festival. This festival provided Constable's first direct experience of Korean percussion and introduction to Kim Juhong and his ensemble Noreum Machi. Building on Synergy's earlier history with Korean ensemble Samul Nori in the 1990s and influenced by the approach to intercultural percussion projects of other Australian percussionists, in 2012 Synergy began to engage with Korean drumming in a manner informed by Taikoz's approach. Initial explorations of Korean percussion commenced through projects with Australian drummer Simon Barker, leading to a series of intensive study trips with Noreum Machi in South Korea. Over a period of four years, the two ensembles exchanged knowledge leading to the intercultural music project titled Earth Cry (2016).

The ongoing engagement of Australian percussionists with Asian percussive practices speaks of Australia's place in the southern hemisphere, and the increasing awareness of this in the late twentieth century. The proximity of Asian countries to Australia began to be acknowledged in the wider cultural landscape during the period in music history that percussion emerged, through the establishment of a range of organisations such as the Asian Studies Association of Australia in 1975. Stanhope suggests that the presence of Asian musical elements in Australian music has 'become a metaphor for saying this is really our place in the world. We are much closer to Asia, therefore we should look to it, listen to it and be influenced by it' (Richards, 2007: 4).

In addition to the prominent influence of Asian musical traditions on Australian percussion music, is the influence of African traditional music. This is unsurprising given that nearly all western percussion instruments can trace historic links to Africa. Williams observes that the development of western percussion aligns with a period in music history when the African diaspora was influencing musical development in North America, which later impacted the evolution of contemporary percussion practices (Williams, 2016: 241). Some of the most well-known twentieth-century Australian percussion ensemble repertoire draws on African traditional music, such as Edwards' *Marimba Dances* (1982), Westlake's *Omphalo Centric Lecture* (1986) and *Fabian Theory* (1987), and Vine's *Defying Gravity* (1987). The use of traditional African marimba ensemble compositional concepts is obvious in the latter three works through their tonality, use of the interlocking rhythmic patterns and direction to performers to share instruments. It is perhaps less obvious in Edwards' *Marimba Dances* (Edwards & Knowles, 1984). This work grew out of the marimba part of *Laikan* (1979), composed for the Fires of London and Sir Peter Maxwell Davies to première at the 1980 Festival of Perth. Although it was commissioned by and is dedicated to the Fires of London percussionist John Gregory Knowles, it was Michael Askill who premièred the solo work. Edwards states that 'the marimba writing is influenced by a transcription I made of music for African harp in my reconstruction of a Madagascan folksong' (Edwards, 2015). A detailed analysis of this work has been produced by the Australian Music Centre (Cooney, 2011). Within ten years of its première, *Marimba Dances* had entered the regularly performed Australian percussion repertoire. It became an especially popular work among students in Sydney with the introduction of NSW Higher School Certificate guidelines mandating inclusion of an Australian work composed in the 25 years prior to any performance exam date. As few attractive marimba pieces of a suitable standard for such exams were available in the 1990s, *Marimba Dances* was regularly performed by all generations of percussionists around the country and subsequently became a reference point for an 'Australian' approach to marimba composition (McCallum, 1990; Vance, 1985). In Western Australia, Paul Tanner's study in Africa is reflected in the tonality and rhythmic structure of frequently performed marimba solos *Old Djenne* (1991), *Her Majesty* (1991) and many others since. The manifestation of influences from the musics of Africa, Asia, Europe and North America in Australian contemporary percussion suggests that active engagement with external influences is a key to delineating the way Australian percussionists engage with their artistic practice. The practice of looking out for inspiration is coupled with a practice of looking in.

Looking in

Landscape and place

It has been suggested by musicologist David Symons that interest in establishing an Australian musical identity via the evocation of landscape in the first half of the twentieth century stems from earlier nationalist trends in British music (Symons, 2016: 10). Literature examining how a sense of place impacts upon the creation of music has been discussed by numerous musicologists, many of whom focus specifically on the relevance of the Australian landscape to Australian composers (Richards, 2007; Kerry, 2009). Investigations into place have begun to appear in percussion research in recent years, in the form of scholarly articles and performance events, such as the 2010 Percussive Arts Society International Convention Focus Day: The Ecology of Percussion. In the same year, the Australian Percussion Gathering Environment Day in Queensland was designed to create 'increased awareness of environmental concerns, refocusing the musician toward the act of listening' (Tomlinson, 2010) and this included percussive improvisations and guided outdoor performances. Any survey of percussion repertoire will show that the use of landscape in a search for identity is not unique among Australian composers, and we need only look to the work of North American composers John Luther Adams and John Cage for proof. Part of what makes the influence of landscape in Australian music unique is the landscape itself, and how it manifests itself in a musical context.

The representation of landscapes as a means of evoking a sense of 'Australianism' became prominent in percussion repertoire in the 1980s via works composed for Synergy Percussion and Nova Ensemble. Kerry observes that when a 'distinctively Australian ambient sound' is referred to in Australian music, it is usually in reference to a work whose musical material is designed to represent the desert or the outback (Kerry, 2009: 20). Although a myriad of different natural or urban Australian landscapes could be used as inspiration for creating an 'Australian' sound, Kerry's observation proves true for two of Synergy's most frequently performed works; Sculthorpe's *Djilile* (1986) and *From Jabiru Dreaming* (1984/1989). The use of indigenous melodies as the basis of the melodic material is frequently cited as a key 'Australianism' of these pieces, and Sculthorpe's use of indigenous melodies has been discussed by a range of scholars (Carmody, 2008; Paget, 2013). That indigenous instruments have not been a major influence in contemporary percussion music by Sculthorpe or his contemporaries can perhaps be attributed to the spiritual significance of Australian indigenous instruments, which are traditionally used primarily to accompany songs and rituals. Further, indigenous Australian instruments are few in number according to ethnomusicologist Alice Moyle, who lists the didgeridoo, song sticks, paired boomerang clappers, the Western Australian rasp and a single-headed cylindrical drum. The drum, found only in North Queensland, is thought to have appeared in the music of this region through contact made with people in Papua New Guinea and the Torres Strait Islands (Moyle, 1991: 26).

A second 'Australianism' evident in the percussion repertoire is the use of notated and improvised soundscapes. Sparse and quiet soundscapes suggestive of the outback or bushland, or sonic impressions of rain and tropical weather reminiscent of the northern wet season are common. Frequently, these soundscapes are complemented by compositional material referencing the activity of insect and bird life in either rhythmic or melodic form. An example can be found in Ross Edwards' *Reflections* (1985). Described by Edwards as 'a delicate, trance-like meditation on intimate minutiae from the sound environment of Central Eastern Australia' that features 'strands of insect sounds' (Edwards, 2015), further analysis of *Reflections* can be found in the liner notes of its CD recording (Hannan, 1989).

Nova Ensemble and Clocked Out frequently use the Australian environment as a point of departure when creating new work and as a way of engaging with their artistic context:

> I like the idea of trying to get the sound of the environment into your pieces, to create a music that is particularly of a place. I do try to get out into the bush and I try to bring that sound in [to the music]. If I can't get my audience out into the environment, then let's bring the environment in to the audience. For example, *Their Kind of Moon* has all those West Australian frogs, which are wonderfully percussive animals. I've done similar things with *Karakamia* using bird song, and a sense of space and dryness, sparseness.
>
> (Pye, 2012)

> Clocked Out exists both as an investigation of percussion and as an investigation of sound in place. . . . I think there are interesting directions one can go to build ideas about sound and percussion in Australia.
>
> (Tomlinson, 2014)

The investigation of percussive sound via engagement with the Australian environment has developed beyond concert repertoire to site-specific improvisations and sound installations. Ros Bandt provides examples of such work created prior to 2000 in *Sound sculpture: Intersections in Sound and Sculpture in Australian Artworks* (Bandt, 2001). In addition to these works percussionists often use site-specific approaches to making new work in order to allow a wider cross section of the community to engage with these ideas. Graeme Leak's *The Musical Fence* (2003), designed and built in Winton, Queensland, is one such example. An inexpensive form of fencing, wire fences stretching vast distances are common in regional and remote Australia. Wire fences are synonymous with an open Australian landscape and thus have been a source of inspiration for other musicians including artists Jon Rose and Hollis Taylor, who have been prolific in their explorations of sounding wire fencing (Taylor, 2007). Also synonymous with the open landscape is the little-known Condamine bell, a non-resonant bell first made by Australian drovers in the 1880s (see Cooney, Knie, & Knie, 2008), featured by Tomlinson and Griswold in the site-specific work *Sounding the Condamine* (2008).

The urban Australian landscape has been increasingly influential in the creation of new percussion work since 2000, reflected in the collection of works composed for the Federation Bells. The Federation Bell project was commissioned by the state government of Victoria through the Melbourne Festival in 1998 to mark Australia's Federation in 2001. Following completion in 2002, a range of Australian composers including Anne Boyd, Brenton Broadstock and Graeme Leak, were subsequently commissioned to make new works using the bells for performance at the site of installation on the banks of Birrarung Marr. The bells have since featured in a number of other works, including Eugene Ughetti's *Bell Curve* (2016).

Influences from urban Australia

Everyday influences on contemporary music as a whole have been present since the early twentieth century. In 1930, Cowell observed that 'contemporary music makes almost universal use of materials formerly considered unusable' and this has proved true particularly within the discipline of percussion (Cowell, 1930: XV). Cage later stated that 'any sound is acceptable to the composer of percussion music; he explores the academically forbidden "non-musical" field of sound insofar as is manually possible' (Cage, 1961: 5). Initial explorations were primarily through everyday items that could be sounded using established percussive techniques of striking or shaking. In the mid-twentieth century the use of everyday items diversified to include objects that required the development of new sounding techniques and performance practices.

The percussive use of everyday items was emphatically embraced by Australian percussionists in the 1980s. This initially grew out of necessity and included repurposing, substitution and instrument building. For Greg Sheehan, Nova Ensemble and AC/PVC, repurposing, substitution and instrument building were originally practical rather than artistic choices, as a range of western and non-western percussion instruments proved difficult to source in Australia at the time. As discussed in Chapter Three, Graeme Leak explored new instrument design using materials close at hand, most of which were inexpensively sourced from the Australian hardware store Bunnings. In the early twenty-first century, Thomas Meadowcroft's works explore a sense of urban Australia in their instrumentation, described by Ughetti as 'the most Australian-sounding percussion music I've ever heard' (Ughetti, 2017b). For example, Meadowcroft's percussion trio *The Great Knot* (2011) reflects the urban environment of the performers, featuring a range of domestic objects such as mixing bowls, tin cans, marbles, table tennis paddles, casio keyboards and the plastic recorders used by children in primary school music classrooms.

In the late twentieth century, Australian percussionists and composers started to reach beyond instrument building towards sculpture and other forms of visual art from a musical perspective. As engagement in this rich area emerged in Australian percussion towards the end of the period of time examined in this book, only a handful of select examples are presented here. A number of early explorations in this area were created for Vanessa Tomlinson, including Rosemary Joy's first percussive sculpture *Puff* (2000). Joy describes this work as 'a large sculpture made of wadding and fabric on a wooden frame with found objects secreted within. The work was intended as a sculptural instrument that had only one possible performance that could be played on it' (Joy, 2015: 13). Anthony Pateras' *Mutant Theatre* (2002) included a display of everyday items that added a sense of excitement to the work: dominoes arranged to trigger mousetraps with balloons attached, light bulbs, marbles and various cheap electronic toys. Although composed for San Diego-based ensemble red fish blue fish [sic], Erik Griswold's kinetic sculpture *Strings Attached* (1999) was performed in Australia shortly after the premiere. It has entered the Australian canon via multiple performances by ensembles around

Australia since Griswold's migration. Part installation, part performance piece, this work requires performers to execute, but is largely focussed on the objects they engage. *Strings Attached* creates a visual representation of sound created by the sticks of six percussionists that are attached to nylon ropes leading either to a lighting rig or to another player, and was followed by *Rope Bridge* (2003), and another kinetic sculpture developed in collaboration with Tomlinson and sculptor Cameron Robbins titled *Spill* (2007). *Spill* features a large cone-shaped pendulum filled with approximately 20 kilograms of rice grains. The performer sonically activates a series of bowls, rice, paper and other objects by placing them under a slow spill of rice falling from the swinging pendulum.

Post-2000, Speak Percussion began collaborating on interdisciplinary works that are performed by percussionists but sit somewhere between installation, instrument design, music and performance art. Although the chronological scope of this study prevents in-depth examination of work in this area, some examples should be mentioned here. An early example is Ughetti and Elaine Miles' *The Glass Percussion Project* (2006–2010), wherein over 2500 hand-blown glass objects were built for two percussionists to sound. More recent examples include *Transducer* (Ughetti & Fox, 2013) and *Fluorophone* (2015). *Transducer* is a through-composed work that 'places the microphone at the centre of the creative investigation and redefines its function and potential as an expressive musical object' (Ughetti, 2017). *Fluorophone* includes five distinct works by a range of composers who each explore light sources as instruments including analogue and digital fluorescent lights, safety matches, strobe lights and custom-built LED discs housed within drums. The influence of the theatre arts on percussion has similarly increased in recent years, foreshadowing greater engagement with percussion theatre in Australian music. The composition of fully notated solo works that use found objects, traditional instruments and elements of theatre has been increasing since 2000. Examples include David Young's *16 Boxes: For Percussionist and Origami Assistant* (2005), Cat Hope's *Sub Aerial* (2015), Ughetti's *Assembly Operation* (2017) and Kate Neal's *Never Tilt Your Chair Back* (2017).

Approach and community

There is no question that the influences discussed earlier are present in contemporary percussion activity in countries other than Australia. Reflected in contemporary percussion music globally are influences from the European avant-garde, traditional musics from around the world and practices developed in the visual and theatre arts. However, it is how these influences come together in the Australian context that was cited by interviewees as integral when considering identity. The responses to questions about the sound of Australian percussion music were related not to compositional devices, repertoire or even particular instruments. Instead, approach to the materials was the central focus. The terms 'community', 'flexibility', 'openness', 'adaptable', 'diverse', 'resourceful' and 'resilient' were used by interviewees both in terms of individual practices and relationship to the wider percussion community. Although it is true that these terms could be used to describe contemporary percussionists in any country, they are significant in the wider Australian geographical and cultural context, as these terms were frequently cited by interviewees in reference to national events designed to draw members of the far-flung Australian percussion community together, such as the 2010 Australian Percussion Gathering.

In part to combat the distance between those working in different hubs, conscious efforts to bring Australian percussionists together with each other and with international artists have been made on a regular basis. The impact of Australian 'percussion gatherings' since 1998 on the practice of significant individuals was repeatedly cited as significant in combating isolation. These events were foreshadowed in the twentieth century by the engagement with international artists including those discussed previously. Australian percussionists marked the start of their own professional contributions to these gatherings with a series of masterclasses in 1998 held at the Australian National Academy of Music, directed by Synergy Percussion with Spanish guest percussionist José Vicente. This and subsequent events served the dual purpose of connecting Australian practitioners with one another, and with pioneering international practitioners. In reflecting on the 2003 Australian Percussion Symposium and Eisteddfod, Tomlinson reflects that it 'demanded that we were in this together, and that was a pretty important moment for me, taking on board the fact that it wasn't just about individual pursuits . . . it was about how one shared those contributions that was important (Tomlinson, 2014). The sense of community support amongst Australian practitioners has enabled fruitful collaborations that have driven artistic change despite the large distances between individuals. Huang observed the impact of this on the development of Australian percussion from an external perspective: 'I have never witnessed this level of collegiality, support, and collaboration in any musical organisation . . . this high level of collegiality is fundamental not only to the survival, but also to the growth of contemporary percussion in Australia' (Huang, 2017: 103).

Also influential on the collective identity of Australian percussion is the transmission of ideas, history and repertoire. Like their international counterparts, the majority of Australian contemporary percussionists learn their craft in conservatoria and universities, albeit with one key difference in the approach. Many of the institutions in Europe and North America where percussion is taught have many decades of western art music education history guiding the institutional environment, with the tradition of orchestral musical training at their core. Australian institutions simply do not have the same weight of local tradition influencing their pedagogical practices. The populations of each major city are small enough that the performing culture at each tertiary institution has an ongoing impact on the professional scene through the percussionists that graduate from their programs. Many of the heads of percussion departments in Australia have had significant influence on professional percussion practices in their respective states, as they are able to design new programs relatively freely to adapt to the musical climate and interests of the time. The result is evident in a generation of percussionists who are willing to attempt anything and do not feel confined creatively. Robbie states:

> Crucial to shaping our percussion identity are the educational institutions and the content on offer to contemporary percussionists. I think in recent times, the content has been adapted and changed altogether; steering slightly away from a purely orchestral approach. Students are viewing current contemporary groups and soloists around the country and are being influenced greatly by what they are hearing.
>
> (Robbie, 2015)

As experimentation is at the heart of all percussion activity, it is true that contemporary percussionists of all nationalities exhibit this quality. However, composers and percussionists have felt that there is a difference in the Australian approach, a difference that has been attributed to the Australian 'have-a-go' mentality. A lack of prior experience or even knowledge of various techniques, repertoire or instruments, is rarely a deterrent:

> One of the things that defines Australian percussionists is that we're all definitely willing to be across a very broad range of instruments and styles . . . we've had to be very inventive to survive in this huge land . . . As a result of that, I think that we are quite malleable as players.
>
> (Ughetti, 2014a)

> I realised how much incredible work there is if you just generate your own scene. One of the great things, I think, about percussion, is that most people are self-reliant because of all of the shit you have to do to be a percussionist.
>
> (Tomlinson, 2014)

> If a short history means there isn't a 'right' way of doing something and you are completely unshackled in your efforts, then perhaps that is a good thing.
>
> (Devenish in Huang 2017: 113)

Tomlinson refers to the Australian percussion approach of 'saying yes, then working out how to do it' (Tomlinson, 2014) when confronted with an unfamiliar task, and notes that some of her long-term collaborative relationships with composers such as Vinko Globokar began this way. A similar observation made by Keith Humble in the early 1970s suggests that this approach has been present since the early days of Australian contemporary percussion activity. When interviewed about the performance of his compositions, Humble stated:

> It is a question of instrumentalists. A critic recently wrote to me saying that he didn't mind what composers were writing, as long as they were writing in a twentieth century idiom. I replied that I had just written a piece for three percussionists, prepared piano and prepared harp. Now, where are the percussion

> players in Australia to perform this piece? I am aware of the percussion works of the last ten years because at the Centre [de Musique in Paris] I had six percussion players – some of the best in Europe. There is a great sense of frustration when you have been working with such performers and you come here and even the material does not exist. But there is a tremendous enthusiasm here – all musicians will 'have a go', and are willing to look at new music. On the contrary, often abroad they will not have a go.
>
> (Murdoch, 1975: 124–125)

Looking forward

In 1975, the number of Australian percussion solo and ensemble works was a single digit number. Forty years later, Australian percussion works numbered in the thousands. Although a handful of these works have been mentioned in this book, there are many others equally worthy of inclusion. The selection of examples cited is in no way designed to bring specific works or methods of approaching percussion music into prominence above others, and as such, these examples appear in no particular order throughout the book. The selected examples herein are included as a means of surveying a range of internal and external influences on twentieth-century Australian percussion practices. Some examination of the influences on the work of Australian percussionists between 1960 and 2010 has been undertaken here in an attempt to understand the context of the work and its role in the evolving practice of contemporary percussion globally. Correspondingly, questions of identity are introduced without intent to pigeonhole or place boundaries around repertoire or ideas, nor does this chapter seek to produce a definitive answer to the question of what Australian percussion music sounds like. An attempt to answer these questions by defining a national approach to percussion is not only difficult, but perhaps unwise given the experimental nature of percussion music making and the extremely stylistically diverse practices around the country. The effects of globalisation on the arts in the late twentieth century include the blurring of boundaries and simultaneous diversification and dilution of influences from a range of art forms. In reflecting on the way identity influences her work, painter Nusra Latif Qureshi observes: 'a layering of identity, to history, where the roots of identity lie . . . my presence as an individual and as a painter determine the course of images I make' (Qureshi in McCaughey, 2014: 346). A layering of external, internal and personal influences is certainly present in the Australian percussion repertoire, but perhaps the more important question is: how does understanding this layering of influences on previous work help or hinder us in producing new artistic work that matters? How does this change Australian perceptions of percussion practices, and how is this viewed from an international perspective?

The previous chapters show that Australian contemporary percussion is constantly evolving, charting an exponential curve of growth since the 1960s. The influences discussed earlier and consideration of identity is perhaps of no consequence to practitioners of contemporary percussion in the twenty-first century, however it cannot be entirely ignored. Stene suggests that questions of identity occupy percussionists globally, albeit for different reasons, and posits this is indicative of a collective identity crisis (Stene, 2015). However, the value of these questions could be the discussion they provoke, particularly as research at the nexus of music and identity is spotlighted in other musicological sub-disciplines. What is clear from the previous chapters is that Australian percussion music has always been more than just Australian. By virtue of the time in history it emerged, Australian percussion music has always been connected to global music activities.

This book does not claim to exhaustively document Australian percussion history, and instead focuses on contemporary percussion activity, with an emphasis on professional percussion ensembles. The primary aim of the book is to serve as an invitation to engage with an art form at the forefront of Australian music. That this documentation is primarily from the perspective of percussionists (rather than composers, musicologists or the media) is particularly important because of the role of performers in the development of percussion practices. Western art music history is littered with the stories of composers and examination of their notated works, while the stories of their performer muses is largely untold. In the field of percussion it is vital to understand the contributions of those who introduced instruments, developed sounding techniques or suggested methods of presentation to composers, and in many cases were composers themselves. It is largely because of the flexibility and creativity of performers that percussion has developed the way that it has, particularly since 2000. Ughetti believes:

> The art form is becoming more discrete in its own right, from the point of view that it can stand over and beyond what general chamber music and contemporary music making is. My analogy would be the way that video art became its own genre in the visual art world, and it's certainly part of my vision to make percussion practice something that stands alone as its own discipline. . . [something that] combines all of these other approaches to the creative process which involve sound design, instrument building and other types of technology.
>
> (Ughetti, 2014b)

Stene questions whether or not percussion will remain an appropriate term for a rapidly evolving practice, while predicting that percussion will 'never refract as an unbroken ray – it will always disperse spectrally' (Stene, 2015: 13). In Australia, a groundswell in practitioner-researchers has taken place since 2000, with interests in technology, digital notation, acoustic ecology and post-percussive practices (see Devenish, 2017). Explorations in contemporary percussion are today pursued by hundreds of specialised players around the country with a variety of approaches. How this will influence the Australian and global development of contemporary percussion practices in the twenty-first century remains to be seen.

References

Atherton, M. 1990. *Australian Made, Australian Played: Handcrafted Musical Instruments from Didjeridu to Synthesiser*. Kensington: New South Wales University Press.

Bandt, R. 2001. *Sound Sculpture: Intersections in Sound and Sculpture in Australian Artworks*. Sydney: Craftsman House.

Cage, J. 1961. *Silence: Lectures and Writings by John Cage*. Middletown: Wesleyan University Press.

Carmody, J. 2008. "The Sound of Other People's Ideas." *Australian Literary Review*, 6 February.

Carnegie, J. 2014. "Music Speaks in Spatially Complex Ways." *The Sydney Morning Herald*, 31 July.

Cooney, D., E. Knie, and P. E. Knie. 2008. *Bells of the Australian Bush*. Toowoomba: Jinglestix.

Cooney, P. 2011. *Dance with Nature: The Chamber Music of Ross Edwards*. Sydney: Australian Music Centre.

Cowell, H. 1930. *New Musical Resources*. New York: Something Else Press.

Cowell, H. 1948. "Current Chronicle: New York." *Musical Quarterly*, 34 (3): 410–415.

Devenish, L., ed. 2017. "Contemporary Percussion Explorations in Twenty-First-Century Australia." *Contemporary Music Review*, 36 (1–2). https://doi.org/10/1080/07494467.2017.1368176

Edwards, R. 2015. "Ross Edwards." www.rossedwards.com

Gottschalk, J. 2016. *Experimental Music since 1970*. New York: Bloomsbury Academic.

Griswold, E., and V. Tomlinson. 2010. "Musical Encounters in the Wide Alley." www.vanessatomlinson.com/writing/MusicalEncountersintheWideAlley.pdf

Hannan, M. F. 1989. "Reflections for Piano and Percussion (Ross Edwards)." In *Anthology of Australian Music on Disc: Handbook*, edited by M. Noone and R. Parker, 11–14. Canberra: Canberra School of Music.

Hartenberger, R., ed. 2016. *The Cambridge Companion to Percussion*. Cambridge: Cambridge University Press.

Huang, A. 2017. "Examining Australian Contemporary Percussion through 'things Uniquely Australian': Interviews with Louise Devenish, Claire Edwardes, and Vanessa Tomlinson." *Contemporary Music Review*, 36 (1–2): 102–117. doi:10.1080/07494467.2017.1368175

Joy, R. 2015. "The Appearance of Sound: Listening to Sculptural Percussion." PhD thesis, Melbourne: Victoria University.

Kerry, G. 2009. *New Classical Music: Composing Australia*. Sydney: New South Wales University Press.

Lewis, K., and G. Aguilar, eds. 2014. *The Modern Percussion Revolution: Journeys of the Progressive Artist*. New York: Routledge.

Logan, C. 2007. "Colin Bright and the Psyche of Place." In *The Soundscapes of Australia: Music, Place and Spirituality*, edited by F. Richards, 35–50. Hampshire: Ashgate.

McCallum, P. 1990. "Synergy Are on the Way to New Music's Golden Fleece." *Sydney Morning Herald*, 15 October.

McCaughey, P. 2014. *Strange Country: Why Australian Painting Matters*. Melbourne: The Miegunyah Press.

Moyle, A. 1991. *Music and Dance in Traditional Aboriginal Culture: Student's Book*. Melbourne: Monash University Faculty of Education.

Murdoch, J. 1975. *Australia's Contemporary Composers*. Melbourne: Sun Books.

O'Neill, L. 2013. "Creating New Music Compositions Based on the Number Systems of Composer Greg Sheehan." Master diss., Wollongong: University of Wollongong.

Paget, J. 2013. "Has Sculthorpe Misappropriated Indigenous Melodies?" *Musicology Australia*, 35 (1): 86–111. doi:10.1080/08145857.2013.761101

Richards, F. 2007. *The Soundscapes of Australia: Music, Place and Spirituality*. Aldershot: Ashgate.

Rogers, V. 2009. *The Music of Peggy Glanville-Hicks*. Farnham: Ashgate.

Sallak, B. 2014. "On the Nature of Percussion Masterworks." In *The Modern Percussion Revolution: Journeys of the Progressive Artist*, edited by K. Lewis and G. Aguilar, 188–198. New York: Routledge.

Schaefer, J. 2007. "More About Michael Askill." www.michaelaskill.com/cms/introduction/34

Schick, S. 2006a. *The Percussionist's Art: Same Bed, Different Dreams*, 1st ed. New York: University of Rochester Press.

Schick, S. 2015. "What Is Good Percussion Playing? Interesting Answers to a Set of Problems (OjaiU2015 class #4.4)." www.youtube.com/watch?v=g35vqXt6dZ4

Schwartz, K. R. 1961. *Minimalists*. London: Phaedon Press.

Stene, H. 2015. "This Is not a Drum." DMA thesis. Norwegian Academy of Music. http://hdl.handle.net/11250/2379520

Symons, D. 2016. *Before and After Corroboree: The Music of John Antill*. Farnham: Ashgate.

Taylor, H. 2007. *Post Impressions: A Travel Book for Tragic Intellectuals*. Portland: Twisted Fiddle.

Tomlinson, V. 2010. *Australian Percussion Gathering*. Mackay: Lanai.

Ughetti, E. 2014a. "Ughetti 5: Rushford – Twin Resistance." In *New Waves*, edited by P. Sametz. Melbourne: Australian Broadcasting Corporation. http://mpegmedia.abc.net.au/classic/australianmusic/201405/amu-2014-05-30.mp3

Ughetti, E. 2017. "TRANSDUCER: Microphones and Loudspeakers as Musical Instruments." *Contemporary Music Review*, 36 (1–2): 48–63. doi:10.1080/07494467

Vance, D. 1985. "Mixed Styles and Talents." *Sydney Morning Herald*, 29 October.

Williams, B. M. 2016. "African Influences on Western Percussion Performance and Pedagogy." In *The Cambridge Companion to Percussion*, edited by H. Russell, 241–252. Cambridge: Cambridge University Press.

Personal communication

Cleworth, I. 2013. Skype Interview with Louise Devenish, 25 March. Sydney, New South Wales.

Constable, T. 2014. Live Interview with Louise Devenish, 4 June. Perth, Western Australia.

France, G. P. 2013. Skype Interview with Louise Devenish, 25 March. Canberra, Australian Capital Territory.

Grigg, A. 2017. Telephone Interview with Louise Devenish, 23 November 2017. Adelaide, South Australia.

Otte, A. 2013. Email Communication with Louise Devenish, 2 July. Cincinnati, Ohio.

Pye, D. 2012. Live Interview with Louise Devenish, 24 September. Fremantle, Western Australia.

Reeves, R. 2015. Live Interview with Louise Devenish, 5 February. Perth, Western Australia.

Robbie, I. 2015. Email Communication with Louise Devenish, 10 December. Perth, Western Australia.

Scholes, L. 2014. Skype Interview with Louise Devenish, 28 June. Melbourne, Victoria.

Talbot, N., and I. Robbie. 2012. Live Interview with Louise Devenish, 10 October. Perth, Western Australia.

Tomlinson, V. 2014. Telephone Interview with Louise Devenish, 5 July. Brisbane, Queensland.

Ughetti, E. 2014b. Skype Interview with Louise Devenish, 27 June. Melbourne, Victoria.
Ughetti, E. 2017b. Personal Communication with Louise Devenish, 4 December. Melbourne, Victoria.
Westlake, N. 2017. Telephone Interview with Louise Devenish, 21 November. Sydney, New South Wales.
White, T. 2014. Live Interview with Louise Devenish, 1 July. Perth, Western Australia.

Bibliography

Works cited

Adamson, J. 1994. "Synergy's One-Night Stand has Reached Its 20th Anniversary." *Sydney Morning Herald*, 28 July.

Amirkhanian, C. 1989. "Lou Harrison and Bill Colvig Interviewed by Charles Amirkhanian." https://archive.org/details/HarrisonColvigInterview

Appleby, R. 2012. *Women of Note: The Rise of Australian Women Composers*. Fremantle: Fremantle Press.

Aronowicz, A. 2014. "Anthony Pateras Alone." *Resonate Magazine*, 17 April. www.australianmusiccentre.com.au/article/anthony-pateras-alone

Askill, M. 1986. *Australian Percussion*. Great Island Records. Compact disc.

Askill, M. 2013b. *A Life in Music*. Personal document. Brisbane: Queensland.

Askill, M. 2015. "Reflections on Composition and Consciousness." PhD thesis, Brisbane: The University of Queensland.

Askill, M. 2017. "Composing with Ancient Sound Technology in the Twenty-First Century." *Contemporary Music Review*, 36 (1–2): 86–101. doi:10.1080/07494467.2017.1368175

Atherton, M. 1990. *Australian Made, Australian Played: Handcrafted Musical Instruments from Didjeridu to Synthesiser*. Kensington: New South Wales University Press.

Bandt, R. 2001. *Sound Sculpture: Intersections in Sound and Sculpture in Australian Artworks*. Sydney: Craftsman House.

Beck, D. 2011. *Set in Stone: The Cell Block Theatre*. Sydney: University of New South Wales Press.

Beck, J. H. 1995. "Percussion Ensembles." In *Encyclopedia of Percussion*, edited by J. H. Beck, 269–273. New York: Garland Publishing Inc.

Bernard, J-P. 2011. "Adventures, New Adventures." www.percussionsdestrasbourg.com

Bessinger, D. 2009. "A Catalog of Works for Marimba Soloist with Percussion Ensemble Composed between 1959 and 2008 with Analysis of Selected Works." DMA diss., Norman: University of Oklahoma.

Bonighton, I. 1977. *Sequenza: And Other Works*. Melbourne: Move Records MS3016–2LP.

Brown, P. 1974. "Memorable 'Rug Concert'." *Sydney Morning Herald*, 12 August.

Burt, W. 2007. "Some Musical and Sociological Aspects of Australian Experimental Music." *Resonate Magazine*, 31 July. www.australianmusiccentre.com.au/article/some-musical-and-sociological-aspects-of-australian-experimental-music

Burt, W. 2017. "Felix Werder and the Accidental Australian Adventure." Accessed December 7 2017. www.dramonline.org/albums/felix-werder-the-tempest/notes

Cage, J. 1961. *Silence: Lectures and Writings by John Cage*. Middletown: Wesleyan University Press.

Cage, J. 1969. *Notations*. New York: Something Else Press.

Calissi, J. L. 2004. "The Marimba Masters." DMA diss., University of North Carolina, Greensboro.

Carmody, J. 2008. "The Sound of Other People's Ideas." *Australian Literary Review*, 6 February.

Carnegie, J. 2014. "Music Speaks in Spatially Complex Ways." *The Sydney Morning Herald*, 31 July.

Cary, T. 1993. "Music: Tristram Cary in Adelaide." *The Australian*, 9 July.

Clark, F. 2014. "TaikOz: More than Muscles in the Media." In *Into the Mix People, Places, Processes: Proceedings of the 2014 IASPM-ANZ Conference*, edited by J. Cattermole, H. Johnson, and O. Wilson, 17–26. Dunedin: International Association for the Study of Popular Music.

Cleworth, I. 2017. "Journeys and Transformations for the Wadaiko in the Australian Cultural Context, Focussing on the Work of TaikOz." Paper presented at the Transplanted Roots Percussion Research Symposium, Brisbane, 23 September.

Cohn, N. 1986. "Ensemble's Rich Percussion Harvest." *The Australian*, n.d. Accessed David Pye's personal collection.

Cohn, N. 1987. "Pitiless Assault by Percussion." *The Australian*, 22 September.

Constable, T. 2017. "Synergy Percussion." Accessed December 7 2017. www.synergypercussion.com

Cooney, D., E. Knie, and P. E. Knie. 2008. *Bells of the Australian Bush*. Toowoomba: Jinglestix.

Cooney, P. 2011. *Dance with Nature: The Chamber Music of Ross Edwards*. Sydney: Australian Music Centre.

Covell, R. 1967. *Australia's Music: Themes of a New Society*. Melbourne: Sun Books.

Covell, R. 1989. "Masterfulness and Doodle: The Healthy Irreconcilables of Australian Music." In *Anthology of Australian Music on Disc: Handbook*. Canberra: Canberra School of Music.

Cowell, H. 1930. *New Musical Resources*. New York: Something Else Press.

Cowell, H. 1948. "Current Chronicle: New York." *Musical Quarterly*, 34 (3): 410–415.

Dalton, G. B. 2006. "Nigel Westlake's 'Omphalo Centric Lecture': A Guide for Performance Including a Biography of the Composer and an Examination of the Different Versions of the Work." DMA diss., Columbus: Ohio State University.

DeHart, J. 2010. "Tap Routes: The Changing Role of the Contemporary Artist-Percussionist." DMA diss., University of California, San Diego.

Devenish, L. 2014. "The Percussion Music of David Pye." *DRUMscene*, Jan/Feb/Mar: 94–96.

Devenish, L. 2015. "And Now for the Noise: Contemporary Percussion in Australia, 1970–2000." DMA thesis, Perth: The University of Western Australia.

Devenish, L., ed. 2017. "Contemporary Percussion Explorations in Twenty-First-Century Australia." *Contemporary Music Review*, 36 (1–2). https://doi.org/10/1080/07494467.2017.1368176

Dunn, R. 1962. *CF Peters John Cage Compositions Catalogue*. New York: Henmar.

Early Warning System. 2017. "About." Accessed 7 December 2017. www.earlywarningsystem.org.au/about

Edwardes, C. 2007. *Coil: Australian Music for Solo Percussion*. Tall Poppies TP193. Compact disc.

Edwardes, C. 2011. *Flash*. Tall Poppies TP215. Compact disc.

Edwardes, C. 2012. "Claire Edwardes Percussion." www.claireedwardes.com/Home.html

Edwardes, C. 2013. *One*. Tall Poppies TP223. Compact disc.

Edwards, R. 2015. "Ross Edwards." www.rossedwards.com

Esler, R. 2014. "I Drum the Body Electric." In *The Modern Percussion Revolution: Journeys of the Progressive Artist*, edited by K. Lewis and G. Aguilar, 112–132. New York: Routledge.

Eyler, D. P. 1985. "The History and Development of the Marimba Ensemble in the United States and Its Current Status in College and University Percussion Programs (Musser)." DMA diss., Baton Rouge: Louisiana State University and Agricultural and Mechanical College.

Eyles, R., and G. Whaley, eds. 1984. *George Hamilton Green's Instruction Course for Xylophone*. Florida: Meredith Music Publications.

Fontana, B. 2017. "Resoundings." Accessed December 7 2017. www.resoundings.org/Pages/Resoundings.html

Fox, R. 2002. "Experimental Music in Melbourne: 1975–1979." Master of Music diss., Melbourne: Monash University.

France, G. P. 1991. "Percussion in Australia." *Percussive Notes*, 29 (4): 23–32.

François, J-C. 1991. "Organization of Scattered Timbral Qualities: A Look at Edgard Varèse's Ionisation." *Perspectives of New Music*, 29 (1): 48–79.

Fulton, A. 2011. "Australian Artists Let in on Ancient Dance." *Sydney Morning Herald*, 25 April. www.smh.com.au/entertainment/dance/australian-artists-let-in-on-ancient-demon-dance-20110424-1dt2s.html

Globokar, V. 1992. "Anti-Badabum." *Percussive Notes*, 31 (1): 77–82.

Gottschalk, J. 2016. *Experimental Music since 1970*. New York: Bloomsbury Academic.

Griswold, E., and V. Tomlinson. 2010. "Musical Encounters in the Wide Alley." www.vanessatomlinson.com/writing/MusicalEncountersintheWideAlley.pdf

Gualda, S., and J-P. Drouet. 1975. *Pour la danse*. LP. Unidisc.

Gualda, S., I. Xenakis, E. Carter, and T. D. Nguyen. 1978. *Percussion Vol 2*. LP. Erato.

Hannan, M. F. 1989. "Reflections for Piano and Percussion (Ross Edwards)." In *Anthology of Australian Music on Disc: Handbook*, edited by M. Noone and R. Parker, 11–14. Canberra: Canberra School of Music.

Hanson, D. 1998. "The History of Sound in the Cinema." *Cinema Technology*, 4: 8–13.

Harley, J. 2004. *Xenakis: His Life in Music*. New York: Routledge.

Hartenberger, R., ed. 2016. *The Cambridge Companion to Percussion*. Cambridge: Cambridge University Press.

Hooper, G. 2012. "Bracing New Music and a New Percussion Ensemble." *RealTime*, 107: 42.

Hooper, M. 2012.*The Music of David Lumsdaine: Kelly Ground to Camberwarra*. London: Routledge.

Huang, A. 2017. "Examining Australian Contemporary Percussion through 'things Uniquely Australian': Interviews with Louise Devenish, Claire Edwardes, and Vanessa Tomlinson." *Contemporary Music Review*, 36 (1–2): 102–117. doi:10.1080/07494467.2017.1368175

Ice, S. P. 2012. "The Percussion Quartet: A Chronological Listing and Performance Guide of Six Selected Works." DMA diss., Norman: University of Oklahoma.

Joy, R. 2015. "The Appearance of Sound: Listening to Sculptural Percussion." PhD thesis, Melbourne: Victoria University.

Kanach, S., ed. 2010. *Performing Xenakis*. New York: Pendragon Press.

Kastner, K. S. 1989. "The Emergence and Evolution of a Generalized Marimba Technique." DMA diss., University of Illinois, Urbana-Champaign. http://hdl.handle.net/2142/21949

Kernan, T. 2014. "What of the Performers." In *The Modern Percussion Revolution: Journeys of the Progressive Artist*, edited by K. Lewis and G. Aguilar, 18–37. New York: Routledge.

Kerry, G. 2009. *New Classical Music: Composing Australia*. Sydney: New South Wales University Press.

Kite, R. 2017. "PAS Hall of Fame: George Gaber." Accessed December 7 2017. www.pas.org/about/hall-of-fame/george-gaber

Knowles, J. 2009. "Setting the Scene: Developments in Australian Experimental Music since the Mid-1990s." In *Experimental Music: Audio Explorations in Australia*, edited by G. Priest, 9–35. Sydney: New South Wales University Press.

Leak, G. 1987. *Programme and Composer Notes: Graeme Leak – Percussion*. Perth: WA Conservatorium of Music.

Leak, G. 2003. *Performance Making*. Sydney: Currency Press.

Leak, G. 2012. "Sketching Impressions: Radio Interview with Eugene Ughetti." *3MBS Radio*, 18 September.

Lewis, K., and G. Aguilar, eds. 2014. *The Modern Percussion Revolution: Journeys of the Progressive Artist*. New York: Routledge.

Logan, C. 2007. "Colin Bright and the Psyche of Place." In *The Soundscapes of Australia: Music, Place and Spirituality*, edited by F. Richards, 35–50. Hampshire: Ashgate.

Ludgate, T., I. Cleworth, M. Askill, and C. Piper. 1989. *Synergy to Paris for French Bicentennial Celebrations*. Sydney: Musica Viva Australia.

Lyons, G. 2011. "Daniel Levitan: An Examination of his Compositional Approach with an Analysis of Marimba Four Hands." DMA diss., Ohio State University. https://search.proquest.com/docview/863656630

Maltese, C. 2011. "A Performance Guide of Selected Works for Horn and Mallet Percussion." DMA diss., University of Miami. https://search.proquest.com/docview/871631484

Martin, C. P. 2017. "Percussionist-Centred Design for Touchscreen Digital Musical Instruments." *Contemporary Music Review*, 36 (1–2): 64–85. https://doi.org/10.1080/07494467.2017.1370794

Martin, P. J. 1996. "Synergy Percussion: Australian Percussion Music in Context." Master of Music thesis, Armidale: University of New England.

McCallum, P. 1990. "Synergy Are on the Way to New Music's Golden Fleece." *Sydney Morning Herald*, 15 October.

McCaughey, P. 2014. *Strange Country: Why Australian Painting Matters*. Melbourne: The Miegunyah Press.

McNeill, R. 2014. *The Australian Symphony from Federation to 1960*. Vermont: Ashgate.

Meyer, J. A. 1999. *Touches of Sweet Harmony: Music in the University of Western Australia 1953–1998*. Nedlands: CIRCME, University of Western Australia.

Moyle, A. 1991. *Music and Dance in Traditional Aboriginal Culture: Student's Book*. Melbourne: Monash University Faculty of Education.

Murdoch, J. 1975. *Australia's Contemporary Composers*. Melbourne: Sun Books.

Neuhaus, M. 1968. *Electronics and Percussion: Five Realisations by Max Neuhaus*. Columbia Masterworks. Compact disc.

O'Connell. 2002. "Father and Son Make Cattle Music." *The Age*, 17 September.

O'Neill, L. 2013. "Creating New Music Compositions based on the Number Systems of Composer Greg Sheehan." Master diss., Wollongong: University of Wollongong.

Otte, A. 1973. "Preferences in Percussion." Michigan State University. www.msu.edu/~sullivan/OttePref.html

Paget, J. 2013. "Has Sculthorpe Misappropriated Indigenous Melodies?" *Musicology Australia*, 35 (1): 86–111. doi:10.1080/08145857.2013.761101

Partch, H. 1974. *Genesis of a Music*, 2nd ed. New York: Da Capo Press.

Pennington, J. 2007. "Lou Harrison's American Gamelan." *Percussive Notes*, 45 (5): 66–68.

Peters, G. B. 1975. *The Drummer Man: A Treatise on Percussion*, Revised ed. Illinois: Kemper-Peters.

Phillips, L. L. 2000. "Vida Chenoweth and Her Contributions to Marimba Performance, Linguistics, and Ethnomusicology." DMA diss., University of North Carolina, Greensboro. https://search.proquest.com/docview/304609811

Piper, C. 2012. "The Percussion Section of the Sydney Symphony Orchestra." *DRUMscene*, 70: 88–92.

Piper, C. 2013. "The Early Days of Synergy." *DRUMscene*, 71: 102–103.

Priest, G. 2009. *Experimental Music: Audio Explorations in Australia*. Sydney: University of New South Wales Press.

Pye, D. 2003. Rebana Loops: Program Note. Accessed November 14 2017. www.novaensemble.com.au/davidpye/rebana.html

Radic, T. 1982. *G.W.L. Marshall-Hall, Portrait of a Lost Crusader: An Introduction to the Marshall-Hall Collection of the Grainger Museum, University of Melbourne*. Perth: Dept. of Music, University of Western Australia.

Richards, F. 2007. *The Soundscapes of Australia: Music, Place and Spirituality*. Aldershot: Ashgate.

Roennfeldt, P. 2011. "Beyond the Ivory Tower – Higher Education Institutions as Cultural Resource: Case Study of the Queensland Conservatorium of Music." *History Compass*, 9: 231–245.

Rogers, V. 2009. *The Music of Peggy Glanville-Hicks*. Farnham: Ashgate.

Roscigno, J. A. 1998. "Percy Graingers's 'The Warriors – Music to an Imaginary Ballet'." Innovations in Orchestration. The Addition of the Melodic Percussion Section as an Equal Force in 20th Century Orchestral Writing." DMA diss., University of Arizona. https://search.proquest.com/docview/304412552

Ryan, R. 2016. "Forging a Singular Musical Path: The Compositional Explorations of Mark Cain." *Sound Scripts*, 5: 33–43.

Sallak, B. 2014. "On the Nature of Percussion Masterworks." In *The Modern Percussion Revolution: Journeys of the Progressive Artist*, edited by K. Lewis and G. Aguilar, 188–198. New York: Routledge.

Schaefer, J. 2007. "More About Michael Askill." www.michaelaskill.com/cms/introduction/34

Schick, S. 1995. "Multiple Percussion." In *Encyclopedia of Percussion*, edited by J. H. Beck, 257–263. New York: Garland Publishing Inc.

Schick, S. 2006a. *The Percussionist's Art: Same Bed, Different Dreams*, 1st ed. New York: University of Rochester Press.

Schick, S. 2006b. "X Is for Xenakis." [liner notes]. In *Xenakis: Percussion Works*. Mode 171/73. Compact disc.

Schick, S. 2014. "Meandering." In *The Modern Percussion Revolution: Journeys of the Progressive Artist*, edited by K. Lewis and G. Aguilar, 201–214. New York: Routledge.

Schick, S. 2015. "What Is Good Percussion Playing? Interesting Answers to a Set of Problems (OjaiU2015 class #4.4)." www.youtube.com/watch?v=g35vqXt6dZ4

Schick, S. 2017. "Keynote Presentation." Paper presented at the Transplanted Roots Percussion Research Symposium, Brisbane, 23 September.

Schmorl, V. 1996. "Australian Percussion Music: An Analytical Study of Selected Percussion Works Commissioned by Synergy." Master of Music diss., Brisbane: Griffith University.

Scholes, L. 2008. "The Evolution of Found Objects in Western Music in the Twentieth Century." Master of Music diss., Melbourne: Victorian College of the Arts.

Schwartz, K. R. 1961. *Minimalists*. London: Phaedon Press.

Silsbury, E. 1996. "A Performance that Goes beyond Music." *The Advertiser*, 30 July.

Skinner, G. 1991. "CD Review: Synergy Percussion." *Sounds Australian*, 31: 46.

Smith, R. 1990. "Skilful Sounds Tailored to Surroundings." *The Advertiser*, 25 August.

Smith, R. 2003. "How the Balloon Taught the Piano to Play." *RealTime*, 55: 28.

Speak Percussion. 2013. "Transducer." http://speakpercussion.com/?page_id=1349&year=32

Stene, H. 2015. "This Is not a Drum." DMA thesis. Norwegian Academy of Music. http://hdl.handle.net/11250/2379520

Stockhausen, K., C. Caskel, M. Neuhaus, and F. Rzewski. 1970. *Cycle for One Percussionist*. Montreal, Canada: Polydor Records.

Sydney Symphony Orchestra. 2017. "Richard Miller: Principal." Accessed December 8 2017. www.sydneysymphony.com/about-us/meet-the-musicians/percussion-and-timpani/richard-miller

Sykes, J. 1973. "Scrounging Nice Sounds." *Sydney Morning Herald*, 24 August.

Symons, D. 2016. *Before and After Corroboree: The Music of John Antill*. Farnham: Ashgate.

Synergy Percussion. 2002. "Steve Reich: Drumming." https://itunes.apple.com/au/album/steve-reich-drumming/id552551262

Synergy Percussion. 2004. *Omphalo Centric Lecture*. Self-published. Compact disc.

Synergy Percussion. 2017. "Synergy: Forty/40." Accessed November 8 2017. www.synergypercussion40under40.com

Taikoz. 2017. "About Taikoz." Accessed December 7 2017. www.taikoz.com/about

Tanner, P. 1995. "Contemporary Percussion Works by Western Australian Composers: An Analysis of Musical Resources and Performance Practice." Master of Music diss., Perth: The University of Western Australia.

Taylor, H. 2007. *Post Impressions: A Travel Book for Tragic Intellectuals*. Portland: Twisted Fiddle.

Tetrafide Percussion. 2004. *Tetrafide Percussion*. Self-published. Compact disc.

Tetrafide Percussion. 2008. *Stick It! Tetrafide Live in Concert*. DVD. Perth: Self-published.

Thomas, A. 1998. "Beware of Snakes, Spiders and Sinclair. John Sinclair (1919–1991), Music Sritic for the Melbourne Herald: The Early Years." *Context: Journal of Music Research*, 15/16: 79–90.

Tomlinson, V. 2010. *Australian Percussion Gathering*. Mackay: Lanai.

Tomlinson, V. 2017a. "On Listening: A Universe of Sound." *Contemporary Music Review*, 36 (1–2): 5–14. doi:10.1080/07494467.2017.1368177

Tomlinson, V., and T. Wren, eds. 2017. *Here and Now: Artistic Research in Music: An Australian Perspective*. Brisbane: Intelligent Arts.

Ughetti, E. 2014. "Ughetti 5: Rushford – Twin Resistance." In *New Waves*, edited by P. Sametz. Melbourne: Australian Broadcasting Corporation. http://mpegmedia.abc.net.au/classic/australianmusic/201405/amu-2014-05-30.mp3

Ughetti, E. 2017. "TRANSDUCER: Microphones and Loudspeakers as Musical Instruments." *Contemporary Music Review*, 36 (1–2): 48–63. doi:10.1080/07494467.2017.1368177

Vance, D. 1985. "Mixed Styles and Talents." *Sydney Morning Herald*, 29 October.

Vickery, L., L. Devenish, S. James, and C. Hope. 2017. "Expanded Percussion Notation in Recent Works by Cat Hope, Stuart James and Lindsay Vickery." *Contemporary Music Review*, 36 (1–2): 15–35. https://doi.org/10.1080/07494467.2017.1371879

Weekes, D. 1996. "Big Noise, Quiet City." *Adelaide Review c.1996*. Accessed personal archives of Vanessa Tomlinson.

Weir, M. P. 2005. "The Role of Vida Chenoweth in Mid-Twentieth-Century Concert Marimba Performance and Literature: A Professional and Personal Perspective." DMA diss., University of Kentucky. https://search.proquest.com/docview/304992277

Werder, F. 1973. *Felix Werder's Banker*. Discovery Stereo. GYS 001. LP recording.

Westlake, N. 1994. *Onomatopoeia*. TP047. Glebe: Tall Poppies. Compact disc.

Westlake, N. 2005. "The Percussion Music of Nigel Westlake: An Interview with Gary France." www.rimshot.com.au/article/the-percussion-music-of-nigel-westlake/

Whiteoak, J. 1989. "Interview with Keith Humble." *NMA*, 7: 21–26. www.rainerlinz.net/NMA/repr/Humble_interview.html

Whitlam, G. 1985. *The Whitlam Government*. Melbourne: Penguin Books.

Williams, B. M. 2016. "African Influences on Western Percussion Performance and Pedagogy." In *The Cambridge Companion to Percussion*, edited by H. Russell, 241–252. Cambridge: Cambridge University Press.

Selected list of musical scores cited

Askill, M. 1990. *Lemurian Dances*. Berlin: E.R.P. Musicverlag.

Blevins, G., and Wilkins, G. 2006. *Columbia Road*. Perth: Self-published.

Bright, C. 1980. *Percussion Quartet*. Sydney: Australian Music Centre.

Brophy, G. 1980. *Hydra*. Sydney: Australian Music Centre.

Burt, W. 1977. *Salon Music*. Melbourne: Self-published.

Cage, J. 1942. *Credo in US*. New York: Peters.

Cage, J. 1960. *27'10.554" for a Percussionist*. New York: Peters.

Cage, J. 1976. *Branches*. New York: Peters.

Carter, E. 1950/1966. *Eight Pieces for Four Timpani*. Minnesota: Hal Leonard.

Edwards, R. 1991. *Prelude and Dragonfly Dance*. Sydney: Australian Music Centre.

Edwards, R., and J. G. Knowles. 1984. *Marimba Dances*. Sydney: Universal Edition.

Edwards, R. 1985. *Reflections*. Sydney: Australian Music Centre.

Feldman, M. 1964. *King of Denmark*. New York: Peters.

Ferneyhough, B. 1991. *Bone Alphabet*. London: Peters.

Griswold, E. 1999. *Strings Attached*. Brisbane: Self-published.

Griswold, E. 2003. *Rope Bridge*. Footscray: Red House Editions.

Griswold, E. 2007. *Spill: Solo Percussion with Kinetic Sculpture*. Brisbane: Self-published.

Humble, K. 1974. *Prime Riff*. Melbourne: Self-published.

Humble, K. 1986. *Sonata for Percussion*. Sydney: Australian Music Centre.

Leak, G. 1984. *. . . And Now for the News*. Melbourne: Self-published.

Nagorcka, R. 1972. *Six Numbers Converging on an Animal*. Sydney: Australian Music Centre.

Pye, D. 2004. *Rebana Loops*. Brisbane: Rhythmscape.

Pye, D. 2007. *Their Kind of Moon*. Perth: Self-published.

Reich, S. 1971. *Drumming*. New York: Boosey & Hawkes.

Rzewski, F. 1985. *To the Earth*. Brussels: Sound Pool Music.

Sculthorpe, P. 1980. *How the Stars Were Made*. Sydney: Australian Music Centre.

Seal, J. 1974. *Structures*. Melbourne: Unpublished.
Stockhausen, K. 1959. *Nr. 9 Zyklus*. London: Universal.
Ughetti, E., and R. Fox. 2013. *Transducer*. Melbourne: Unpublished.
Varèse, E. 1931. *Ionisation*. New York: Ricordi.
Vine, C. 1987. *Defying Gravity*. Sydney: Chester Music.
Werder, F. 1971. *Oscussion*. Melbourne: Self-published.
Westlake, N. 1984. *Omphalo Centric Lecture*. Broadway, NSW: Australian Music Centre.
Westlake, N. 2004. *Kalabash*. Turramurra, NSW: Rimshot Music.
Xenakis, I. 1975. *Psappha*. Paris: Editions Salabert.
Young, D. 2005. *Sixteen Boxes: For Percussionist and Assistant Origamist*. Sydney: Australian Music Centre.

Personal communication

Askill, M. 2013. Skype Interview with Louise Devenish, 7 February. Mullimbimby, New South Wales.
Bailey, J. 2012. Skype Interview with Louise Devenish, 10 November. Penang, Malaysia.
Cain, M. 2013. Email Communication with Louise Devenish, 3 December.
Cleworth, I. 2013. Skype Interview with Louise Devenish, 25 March. Sydney, New South Wales.
Constable, T. 2014. Live Interview with Louise Devenish, 4 June. Perth, Western Australia.
Cossom, R. 2014. Telephone Interview with Louise Devenish, 1 March. Melbourne, Victoria.
Couch, W. 2014. Live interview with Louise Devenish, 10 November. Melbourne, Victoria.
Dovey, R. 2014. Email communication with Louise Devenish, 13 March.
Edwardes, C. 2013a. Live Interview with Louise Devenish, 23 April. Sydney, New South Wales.
France, G. P. 2013. Skype Interview with Louise Devenish, 25 March. Canberra, Australian Capital Territory.
François, J-C. 2012. Email Interview with Louise Devenish. Paris, France.
Green, F. 2017. Telephone Interview with Louise Devenish, 15 November. Adelaide, South Australia.
Grigg, A. 2017. Telephone Interview with Louise Devenish, 23 November. Adelaide, South Australia.
Johnson, D. 2013. Telephone Interview with Louise Devenish, 28 November. Adelaide, South Australia.
Johnson, D., and R. Pusz. 2014. Live Conversation with Louise Devenish, 6 March. Adelaide, South Australia.
Lagos, R. 2013. Live Interview with Louise Devenish, 25 April. Sydney, New South Wales.
Leak, G. 2013. Skype Interview with Louise Devenish, 16 July. Melbourne.
Leak, G. 2014. Email communication with Louise Devenish, 24 August.
Lloyd, R. 2013. Email communication with Louise Devenish, 17 December. Adelaide, South Australia.
Montgomery, D. 2014. Skype Interview with Louise Devenish, 26 March. Brisbane, Queensland.
Neville, P. 2013. Live Interview with Louise Devenish, 31 July. Melbourne, Victoria.
Oates, C. 2017. Email communication with Louise Devenish, 2 December. Adelaide, South Australia.

O'Kelly, T. 2013. Live Interview with Louise Devenish, 22 November. Brisbane, Queensland.
Otte, A. 2013. Email Communication with Louise Devenish, 2 July. Cincinnati, Ohio.
Piper, C. 2013b. Live Interview with Louise Devenish, 26 April. Sydney, New South Wales.
Pratt, D., and R. Miller. 2016. Skye Interview with Louise Devenish, 27 October. Sydney, New South Wales.
Pusz, R. 2013. Skype Interview with Louise Devenish, 13 May. Adelaide, South Australia.
Pusz, R. 2014. Email Communication with Louise Devenish, 27 March.
Pye, D. 2012. Live Interview with Louise Devenish, 24 September. Fremantle, Western Australia.
Quinn, B. 2013. Live Interview with Louise Devenish, 23 July. Moorabbin, Victoria.
Reeves, R. 2015. Live Interview with Louise Devenish, 5 February. Perth, Western Australia.
Robbie, I. 2015. Email Communication with Louise Devenish, 10 December. Perth, Western Australia.
Scholes, L. 2014. Skype Interview with Louise Devenish, 28 June. Melbourne, Victoria.
Seal, J. 2012. Skype Interview with Louise Devenish, 18 December. Melbourne, Victoria.
Seal, J. 2013. Email Communication with Louise Devenish, 20 March.
Talbot, N., and Robbie, I. 2012. Live Interview with Louise Devenish, 10 October. Perth, Western Australia.
Tanner, P. 2012. Live Interview with Louise Devenish, 27 September. Perth, Western Australia.
Timcke, A. 2013. Live Interview with Louise Devenish, 24 July. Perth, Western Australia.
Tomlinson, V. 2014. Telephone Interview with Louise Devenish, 5 July. Brisbane, Queensland.
Tomlinson, V. 2017b. Personal Communication with Louise Devenish, 10 November.
Ughetti, E. 2014. Skype Interview with Louise Devenish, 27 June. Melbourne, Victoria.
Ughetti, E. 2017. Personal Communication with Louise Devenish, 4 December. Melbourne, Victoria.
Wain, G. 2013. Email Communication with Louise Devenish, 17 April.
White, T. 2012. Live Interview with Louise Devenish, 9 October. Perth, Western Australia.
White, T. 2014. Live Interview with Louise Devenish, 1 July. Perth, Western Australia.

Index

Note: *Italicized* page numbers indicate a figure on the corresponding page. Page numbers in **bold** indicate a table on the corresponding page.